Cambridge Elements ⁼

Elements in the Philosophy of Biology
edited by
Grant Ramsey
KU Leuven
Michael Ruse
Florida State University

MODEL ORGANISMS

Rachel A. Ankeny
University of Adelaide

Sabina Leonelli
University of Exeter

CAMBRIDGE
UNIVERSITY PRESS

CAMBRIDGE
UNIVERSITY PRESS

University Printing House, Cambridge CB2 8BS, United Kingdom

One Liberty Plaza, 20th Floor, New York, NY 10006, USA

477 Williamstown Road, Port Melbourne, VIC 3207, Australia

314–321, 3rd Floor, Plot 3, Splendor Forum, Jasola District Centre,
New Delhi – 110025, India

79 Anson Road, #06–04/06, Singapore 079906

Cambridge University Press is part of the University of Cambridge.

It furthers the University's mission by disseminating knowledge in the pursuit of
education, learning, and research at the highest international levels of excellence.

www.cambridge.org
Information on this title: www.cambridge.org/9781108742320
DOI: 10.1017/9781108593014

First published 2020

A catalogue record for this publication is available from the British Library.

ISBN 978-1-108-74232-0 Paperback
ISSN 2515-1126 (online)
ISSN 2515-1118 (print)

Model Organisms

Elements in the Philosophy of Biology

DOI: 10.1017/9781108593014
First published online: October 2020

Rachel A. Ankeny
University of Adelaide

Sabina Leonelli
University of Exeter

Author for correspondence: Rachel A. Ankeny rachel.ankeny@adelaide.edu.au

Sabina Leonelli S.Leonelli@exeter.ac.uk

Abstract: This Element presents a philosophical exploration of the concept of the 'model organism' in contemporary biology. Thinking about model organisms enables us to examine how living organisms have been brought into the laboratory and used to gain a better understanding of biology and to explore the research practices, commitments, and norms underlying this understanding. We contend that model organisms are key components of a distinctive way of doing research. We focus on what makes model organisms an important type of model and how the use of these models has shaped biological knowledge, including how model organisms represent, how they are used as tools for intervention, and how the representational commitments linked to their use as models affect the research practices associated with them. This title is also available as Open Access on Cambridge Core.

Keywords: models, representation, experimentation, abstraction, research practices

ISBNs: 9781108742320 (PB), 9781108593014 (OC)
ISSNs: 2515-1126 (online), 2515-1118 (print)

Contents

1 Model Organisms

1.1 Introduction

This Element presents a philosophical exploration of the concept of the 'model organism' in contemporary biology. Thinking about model organisms enables us to examine how living organisms have been brought into the laboratory and used to gain a better understanding of biology, and to explore the research practices, commitments, and norms that have made such understanding possible.[1]

We contend that model organisms are key components of a distinctive way of doing research. This way of doing research parallels broader trends in contemporary biology, including moves towards 'big science' approaches, particularly in relationship to the large-scale genomic sequencing projects of the 1990s. It also is unique due to its emphasis on projecting data beyond their original domain and establishing their broader applicability, especially to questions relating to human health and disease. We focus on what makes model organisms an important type of model within the contemporary life sciences, and how the use of these models shapes biological knowledge.

The Element is thus centred on six sets of interrelated questions. First, what do model organisms represent? How does this role compare to others that organisms play in biological research, and in particular how does it relate to non-representational functions of model organisms (such as their use as tools for intervention)? Second, how do model organisms represent, and how do processes of idealisation and abstraction contribute to and warrant the use of such organisms? Third, for whom do they represent? What is the relationship between such organisms and the experimental contexts within which they are utilised? How do the epistemic structures and shared scientific practices within the communities of scientists focused on these organisms influence the ways in which research is conducted and how these organisms are understood? Fourth, why are model organisms accepted as credible representations of biological phenomena? When and why are arguments about projectability of data and other results well founded? Fifth, in what sense can model organisms be thought of as a scientific model? How is knowledge created using these models, and how do the representational and interventionist roles of these models intersect? Finally, what is likely to be the legacy of these models, and what scientific roles are they likely to play in the future?

---

[1] For a systematic analysis of the role of commitments and modelling in achieving scientific understanding, see Leonelli (2009).

1.2 What Are Model Organisms?

Particularly since the advent of large-scale genomic sequencing associated with the international Human Genome Project (HGP), the term 'model organism' has become ubiquitous in contemporary biological discourse. It is difficult to trace the precise point in history at which the actual term was introduced. Use of aspects of the underlying concept can be traced back within organism-based research programmes, particularly in the 1960s and 1970s, in part due to the rise of the techniques associated with molecular biology. Formalisation of the terminology occurred in the 1990s via the HGP which, in turn, resulted in increased numbers of publications associated with certain organisms (Dietrich, Chen, & Ankeny 2014). In the most general terms, model organisms are non-human species that are extensively studied in order to understand a range of biological phenomena. The hope is that data and theories generated through use of the model will be applicable to other organisms, particularly those that are in some way more complex than the original model, especially humans. The most widely acknowledged inventory of these organisms includes those officially recognised by the US National Institutes of Health (NIH 1999) as model organisms for biomedical research, which ultimately listed thirteen species, including mouse (*Mus musculus*), rat (*Rattus norvegicus*), zebrafish (*Danio rerio*), fruit fly (*Drosophila melanogaster*), nematode (*Caenorhabditis elegans*), baker's yeast (*Saccharomyces cerevisiae*), and thale cress (*Arabidopsis thaliana*).[2]

Much biological research aims at extrapolating knowledge beyond the organisms that are actually being studied. The study of an individual specimen is taken to provide understanding about all other members of the same species. Further, it is often expected that the study of a single species will provide biological insights into many other species, though this type of claim is necessarily contingent and requires empirical justification in any particular case. This idea is grounded in evolutionary theory, according to which all life forms are related through a common evolutionary history and thus share a smaller or greater amount of genetic make-up and a number of developmental features. Evolutionary or phylogenetic conservation can be both genomic and also developmental and mechanistic (Love & Trevisano 2013). It is therefore used to justify the treatment of an organism as a sample of a larger class of organisms that are phylogenetically related to that species and hence display significant

---

[2] We generally provide both common and scientific names for organisms, but also use genus only, for instance when a range of species within a genus is being studied, and genus only or common names with reference to the usual designator associated for a particular community of researchers (e.g., the worm or the *Arabidopsis* community), or otherwise where more appropriate.

morphological, structural, developmental, or other types of similarities with it (Weber 2005). As a consequence, a model organism can represent other species: we discuss precisely how this representational role works in more detail in Section 2.

Model organisms have a variety of well-recognised experimental and pragmatic advantages. For instance, they are typically easy to breed and maintain in large numbers under laboratory conditions. Model organism research characteristically involves the standardisation of the organism in question and the accumulation of knowledge and resources on the organism on a large scale. These resources include relevant networks, organism-focused conferences, stock centres, and cyberinfrastructures. Such research is done with a view to creating a platform for interdisciplinary integration across biological disciplines and a reference point for comparative research across species. Many features of model organisms are thus the result of human interventions, including domestication. One of their main functions is to support scientific and technological interactions with the biological world. Indeed, a well-recognised characteristic of model organisms is their usefulness as tools for biological intervention, for example when they are employed to investigate and test techniques for genetic modification or phenotypic imaging.

In this Element, we contend that the representational and interventionist roles of model organisms are deeply linked. Even in situations where model organisms are used primarily as tools to intervene in the biological world, the representational commitments associated with this type of modelling (which we discuss subsequently) persist and underpin their use in research practices. In other words, we argue that there can be no adoption of hybridisation probes or gene-mapping techniques developed on model organisms without also making the representational commitments involved in using those organisms as models. These commitments have thus become entrenched in biology in ways that are often difficult to challenge, despite novel findings that draw the representativeness of model organisms into question, such as gene–environment interactions.

1.3 The Significance of a Label

The term 'model organism' has come to serve not only as a descriptor for organisms used in biological research that have certain attributes but also as a label with prescriptive power. Large amounts of attention and funding were poured into model organism–related research in the 1980–2010s, with the HGP sequencing efforts providing a crucial incentive and rallying point for the need to focus on a limited number of species. Partly as a consequence of such investments, model organisms have played the role of reference point or touchstone for a wide variety of research questions and approaches to biological practice (Ankeny & Leonelli 2011). Some critics have argued that the model organism concept is 'swamping out' contemporary biological research agendas, particularly in terms of funding, making it difficult to pursue biological research on organisms not considered to be official 'model organisms' and to use techniques and methods that do not include or prioritise molecular approaches (Bolker 1995; Davies 2007), although empirical data on publication patterns do not tend to support these claims (Dietrich, Chen, & Ankeny 2014). Others have criticised what they have termed 'organismism', namely over-reliance on model organisms without sufficient attention to whether particular organism-based models are adequate (Robert 2008).

Given the significance of the label, questions about whether and in what sense a particular experimental organism is a 'model organism' require explorations that go beyond abstract philosophical analyses or laboratory boundaries; answers to these questions have clear epistemic, social, political, and economic implications with regard to how science is conducted and how knowledge is constructed. Hence, we provide a philosophical examination of the model organism concept that is grounded in the extensive body of previous scholarly work on relevant contemporary and historic scientific practices in the biological and biomedical sciences (for a detailed historiographic overview of this literature, see Ankeny & Leonelli 2018). This analysis makes a critical contribution to the literature on the philosophy of biology and has important implications for the conduct of contemporary science, including how we understand the underlying epistemic structures and scientific practices relating to this type of research.

1.4 Grounding Philosophy in the Study of Research Practices

This Element identifies and analyses philosophical issues associated with the concept of a 'model organism' against the backdrop of in-depth empirical study of the history of these organisms within biology and the practices within the fields associated with this type of research (although we do not develop any detailed historical accounts about particular organisms in this context). We therefore intertwine descriptive and normative analysis of scientific practices in developing and presenting our account. This approach is necessary because understanding how model organisms work as scientific models involves understanding how scientists use them in their everyday work and reasoning practices, and how those uses and associated arguments have changed over time.

This type of scientific practice cannot be documented using published articles alone, which typically provide a line of reasoning and a set of conclusions without reporting all of the processes through which these were developed. Given the limits of published literature for analysing how organisms are actually used, it was essential for our account that we develop and utilise a range of rich descriptive materials using techniques from history and the social sciences. In this Element, we rely on previously published scholarly literature, archival material, scientific records (such as grant applications and institutional records), and grey literature (such as newsletters, reports, guidelines, 'how-to' documents released by stock centres, and databases). We also carried out interviews with researchers and others involved in scientific practice. These included administrators and technicians from various labs, at different career stages, from diverse fields, and working in different geographic locations. Additionally, we made ethnographic visits to observe practices in laboratories, field sites, funding institutions, scientific conferences, and other settings.[3] Finally, we draw on our long-term collaborations with practising scientists through common projects and publications, membership of expert working groups, and advisory positions in steering committees and stock centres relating to model organism biology. These activities have increased our exposure to laboratory life both at the policy and organisational level – through the perspectives of relevant funders, learned societies, and institutions – and at the level of researchers' own interests and strategies, including the constraints and impediments that they face.

This philosophical study of model organisms thus exemplifies the value and importance of fostering collaboration between humanists and scientists, as well as constructive dialogues across subfields that focus on the contemporary life sciences within the history and philosophy of science (HPS) and science and technology studies (STS). Understanding how an organism can function as a scientific model means delving into questions concerning the value, epistemic pay-offs, and skills involved in manipulating a physical object (rather than a mathematical construct or a simulation). It also requires reflection on ways that the relationships between researchers and organisms, which include familiarity and affect, may shape biological understanding and resulting knowledge. Equally critical are the roles played by instruments, techniques, institutions, and infrastructures organised around the organism in channelling and entrenching particular ways of doing research. Finally, it is important to consider the extent

---

[3] For some of the archived materials on which this Element draws, see the Zenodo data collection 'Exeter Data Studies' (https://zenodo.org/communities/datastudies) which includes interview materials with researchers who work on *Arabidopsis and* various types of yeast; the Bermuda Principles data archive which includes interviews relating to the model organism projects within the Human Genome Project, housed within the DukeSpace Archival Collections, Center for Public Genomics Research Files (https://dukespace.lib.duke.edu/dspace/handle/10161/7407); and the Organisms and Us website at https://arts.adelaide.edu.au/organisms-and-us/.

to which historically rooted commitments and social dynamics contribute to the development, use, and interpretation of these models. Far from a matter of logical reconstruction informed solely by the study of scientific publications, elucidating the epistemic role of model organisms within biology requires situating these research components in their material, social, and historical contexts.

2 What Do Model Organisms Represent?

2.1 Introduction

Model organisms help to create knowledge that can be projected beyond the immediate domain in which it was produced. We argue that this projection happens simultaneously in two respects: in terms of the range of organisms being represented (what we call 'representational scope') and the type of phenomena that model organisms are used to study ('representational target'). We then consider the implications of this claim for understanding the representational power of model organisms as scientific models and comparing it to other ways in which organisms are used and interpreted within research. This account emphasises the characteristics associated with model organisms that are necessary to ground their abilities to serve as models (but does not yet address the question of what makes a *good* model, which we confront in Section 5). These characteristics are simultaneously biological and epistemic, and are shared by all model organisms to a greater or lesser extent.

2.2 Representational Scope

Why do biologists study fruit flies, worms, or mice, when they are actually interested in humans or biological processes in general? Some species may well be of interest to biologists in and of themselves. But when specific organisms are selected and studied as model organisms, researchers are typically claiming that they will provide some information or understanding about forms of life beyond the original focal organism. We use the term 'representational scope' to describe how extensively the results of research conducted on a group of specimens (tokens) can be projected onto a wider group of organisms labelled through reference to a type (e.g., a taxonomic class), a classic form of the problem of induction. The projection can vary from a single species for which the organism is serving as a proxy (notably humans) to a wider class of organisms such as a family or a kingdom (say all mammals or animals), or perhaps even to all organisms, if a process or phenomenon is thought to be universal or common. The extent of representational scope assumed by researchers is often related to the criteria for the selection of the organism in the first instance, together with the particular context of use and the questions or processes to be investigated. This concept is a critical epistemological feature that shapes which organisms are selected as a research focus and how they are developed for research.

The representational scope of an experimental organism can be very narrow and extend only to its own species or those that are closely related: for instance, red-eared terrapins are used to study turtle shell development (Maher 2009) and tamar wallabies are used as a model for reproduction and development in kangaroos, and marsupials more generally (Hickford, Frankenberg, & Renfree 2009). Researchers may hope that the study of these organisms reveals something about behaviour or physiology that is generalisable. However, this outcome is rarely attained, particularly for research that does not rely on previous empirical evidence about evolutionary or phylogenetic conservation.

By contrast, model organism research programmes share an underlying interpretation of the representational scope of their organisms; the assumed or hypothesised representational scope is broader and more inclusive in the case of model organisms than the representational scope assigned to other experimental organisms. It is common for the results of *D. melanogaster* genetics or *C. elegans* physiology, for example, to be interpreted as applying to a much wider range of organisms, often including humans. Even in the cases of *A. thaliana* and *S. cerevisiae*, findings have been projected well beyond the realm of plants and fungi respectively.

Model organisms serve as the basis for articulating processes that are thought to be common across all (or most) other types of organisms, particularly those processes whose molecular bases can be articulated. Hence, it is often claimed that processes in model organisms are representative of processes shared by higher level organisms, especially humans: in other words, 'the fish is a frog . . . is a chicken . . . is a mouse' (Kimmel 1989, as paraphrased in Grunwald & Eisen 2002, 721). The most common sense in which these organisms are 'representative' relates to their use in the HGP and, in most cases, as models which provide the basis for biomedical research. Model organisms thus lie at one extreme of the spectrum associated with representational scope, namely being associated with a high degree of generalisability.

2.3 Representational Target

Another sense in which findings from research on organisms can be generalised is the number and type of phenomena to which organisms allow experimental access. What are being studied using model organisms? We utilise the term 'representational target' to indicate the collection of phenomena that are to be explored through the use of an organism. By 'phenomena', we refer to the labels used by researchers to define concepts, entities, and processes related to their research interests. Whether understood as observable or unobservable, deeply theory-laden, or 'mirroring' reality in an objective way, phenomena constitute for us the object of scientific claims. Thus, anything from 'metabolism' to a 'hox gene' constitutes a phenomenon and can become the representational target (cf. Meunier 2012).

What is epistemologically distinct about model organisms is their representational target: they serve as models for a relatively wide range of systems and processes that occur in living organisms, including those studied within genetics, development, physiology, evolution, and ecology. This approach allows pursuit of one key goal associated with this type of research: to perform large-scale, comparative work across species, integrating a range of disciplinary research approaches. This goal is achieved using a specific strategy, initially gathering resources and building infrastructure on individual whole organisms, and integrating a range of disciplinary approaches, followed by work on comparisons between these organisms using the model organism as a reference point. For example, a number of homologous genes have been identified across a range of model organisms. Researchers conceptualise identification of these homologs as a key step in producing knowledge about the molecular basis of phenotypes across very different types of organisms, and particularly of variations associated with disease (e.g., the gene *BRCA1*, which is associated with human breast cancer and whose homolog has been found in variant forms in *C. elegans* and *M. musculus)*.

Another example of the fruits of such a research strategy can be found in the elucidation of the mechanisms associated with programmed cell death, which is a regulated process that generally confers some sort of advantage during an organism's life cycle. Using the nematode worm *C. elegans*, researchers identified key genes regulating the processes of cell death in this organism (for a summary, see Wood et al. 1988). It was subsequently shown that corresponding homologous genes exist in higher species, including human beings, and that the basic morphological and biochemical features of programmed cell death are conserved in both the plant and animal kingdoms. In these sorts of research programmes, understanding molecular and developmental processes in the model organism is the initial focus of research which then serves as a building block or platform (e.g., *C. elegans* Sequencing Consortium 1998) for a more general investigation of developmental processes together with molecular and other processes across a much wider range of organisms.

What, then, distinguishes model organisms from the general class of experimental organisms in terms of their targets? The difference does not lie solely in the capacity of these organisms to support human interventions or in their use as tools in research practice; all experimental organisms are, to a greater or lesser extent, used as scaffolds for developing techniques for the control and manipulation of biological processes. Rather, what defines model organisms as a specific subclass of experimental organisms is the representational power attributed to them. This representational power is in turn grounded in the specific modes of intervention and standardisation used to establish and develop these organisms over the past few decades.

Model organisms explicitly represent whole organisms; they simultaneously allow access to specific processes and are investigated using a range of disciplinary approaches with the intention of integrating these approaches to develop a multi-level understanding of their evolution, structures, and behaviours. In contrast, experimental organisms are models for specific phenomena, to be investigated through a particular discipline or approach with its accompanying set of techniques and practices. Thus, experimental organisms need not be as versatile as model organisms in order to be useful and successful for particular types of research. For instance, even if it would be extremely difficult to study dogs in genetic terms due to their relatively large genome size and long generation times, these limitations make them no less valuable for the study of behaviour or disease.

We should note that while mechanisms are clearly an important target for many explanations derived from research with model organisms, and biologists place high value on elucidating mechanisms as an epistemic goal, we do not view mechanistic reasoning as the only type of reasoning associated with model organism research. Precisely due to the emphasis on multi-level integration, causal-mechanistic approaches are combined with mathematical models and simulations of dynamic processes both within and beyond the cellular scale (e.g., intercellular transport and protein folding: see O'Malley et al. 2014). Moreover, understandings of gene functions have benefitted from increasingly data-intensive analysis of the correlations between metabolic and gene expression profiling and phenotypic differences across specimens, which may well underpin causal reasoning but do not necessarily involve the formalisation of mechanisms or even a molecular gene concept (Waters 2013). Since we do not take mechanistic reasoning as the sole goal or the primary means of model organism research, we will not delve here into related philosophical debates on causal reasoning and reductionism, which have been well covered in the existing philosophical literature (for a summary, see Brigandt & Love 2017).

2.4 What Is Represented: The Whole Organism *and* Other Organisms

In our view, the distinctive representational power of model organisms stems from the simultaneous attribution of wide representational scope and wide representational target. They are at the same time models of (many) higher organisms, thus instantiating properties common to many other species, *and* models of the complex interrelations of processes and entities that occur in and make a whole organism, thus instantiating the interdependencies and links between different biological phenomena and diverse levels of analysis.

By contrast, consider Jessica Bolker's account (2009), which distinguishes two types of animal models: what she calls 'exemplars' (or 'proxies'), which are examples of a larger group such as a taxon or other more extensive groups, and 'surrogates', which are substitutes for another entity of special interest, particularly humans in the biomedical sciences. She stresses that when researchers take elucidation of shared fundamental patterns as their aim, organisms are used as exemplary models; this type of goal would be present in most model organism work, and most often occurs in 'basic' research (in our view, any stark, principled distinction between 'basic' and 'applied' or even 'translational' research is difficult to maintain, but detailed discussion of this issue is not necessary for our current purposes). In contrast, Bolker maintains that organisms used as surrogates are substituted for what would be the ideal target (in many cases humans) for ethical or pragmatic reasons, but that those using such models do not necessarily seek to understand underlying biological processes or mechanisms since this is not necessarily required to develop applications such as medical treatments and therapies.

This way of distinguishing the functions served by various animal models does not apply cleanly to model organisms, even though it may initially appear that Bolker's categories can be directly mapped onto our distinction between representational scope and representational target. Her notion of a 'surrogate model' exemplifies a very specific type of representational target (one that has a clear translational role and is most commonly associated to biomedical research on rodents, as we discuss in 4.6), and one that simultaneously implies a limited representational scope. Taking a wider spectrum of model organisms into account, and particularly the common features characterising thale cress (*A. thaliana*), fruit fly (*D. melanogaster*), nematode (*C. elegans*), baker's yeast (*S. cerevisiae*), and zebrafish (*D. rerio*), we contend instead that both the target and the scope of model organisms are typically broad, and that focus on projecting results across a wide range of species does not diminish researchers' interest in targets including molecular, developmental, and evolutionary processes (and their interactions). In other words, model organisms can and often do serve both as exemplars and surrogates, though this way of framing their representational role takes attention away from the broader range of species for which they function as models (i.e., beyond the human).

Another account that only partially captures what we take to characterise model organisms is the one provided by Arnon Levy and Adrian Currie (2015; see also Parkkinen 2017), which stresses the importance of shared ancestry as a unique part of what makes model organisms a distinct type of model. They argue that inferences using them rely on empirical extrapolations such that biologists can treat the organism as a representative specimen of a broader class, which in turn is part of a more general biological strategy known as the comparative method. As discussed in more detail later (2.4), we agree that commitments to working assumptions about shared ancestry and genomic and other forms of evolutionary conservation are an essential feature of model organism research, but they are certainly not the sole component of what makes model organisms a distinct type of model. Again, we contend that it is the simultaneous attribution of a wide representational scope and a wide representational target, together with attention to how these features intersect and are instantiated in research practice, which make model organisms unique models that represent and are used in a distinct way.

Michael Weisberg (2013) explores what he terms 'model organisms' by using a very wide initial definition. This definition includes not only the canonical model organisms on which we focus our account, but also any organisms used as models in any sense to study something beyond themselves, including humans as well as broad classes of phenomena (such as the use of rabbits in Australia to study invasive species in ecology). This broad definition allows him to frame model organisms as one subset of the broader class of concrete models whose representational power stems from their resemblance to a concrete target. Thus according to him, model organisms differ from other models only insofar as they are not constructed and have their origins in the wild: there are no other special properties of relevance. We disagree with this account in numerous ways. His views rely on a definition of model organisms that is empirically imprecise and overly inclusive, thus missing the significant features of these models and making it impossible to explain their central role in biology within the last century. As we explain later, though of course model organisms have their origins in the wild, they are in fact constructed through a diverse range of practices. These have in turn facilitated their adoption as reference tools to develop techniques and problems for biological, and especially genetic, manipulation. Moreover, the simultaneous focus on a broad representational target and a broad representational scope does constitute a special property of traditional model organisms. It differentiates their use as models from other uses of organisms in research and makes them a special case within the broader category of material models (see also Frigg & Hartmann 2018). We do not contend that they are the only scientific models that have these attributes (this may well be the case, but is not significant for our argument), but that they are particularly notable and important examples of this type of focus.

Finally, it is important to note that our analysis does not map easily onto distinctions used by Mary Morgan or Evelyn Fox Keller with regard to the representational functions of organisms. They both draw a distinction between the notions of 'representative of' versus 'representative for'. In Morgan's account (2003, p. 230; see also her 2007 and Ratti 2018), the distinction captures a difference in the scope of the representation: 'representative of' indicates a narrow, endogenous scope, while 'representative for' stands for broad exogenous scope (e.g., the laboratory mouse *M. musculus* is representative of mice and may be representative for humans). This distinction is compatible with our account with regard to the concept of 'representational scope' but it does not apply to what we call the representational target. Keller's account (2000) differs from Morgan's insofar as it focuses on the purposes for which a model is used, which she calls 'representative for', as opposed to being 'representative of' specific phenomena. While Keller's account usefully places emphasis on the epistemic role of the goals of representation in the case of experimental organisms, it again does not capture the difference between the target and the scope of the model. We believe this distinction is crucial for understanding the epistemic functions of various types of research organisms, particularly model organisms.

While benefitting from dialogue with the above-mentioned scholars, our account is most obviously complementary to philosophical views on model organisms that are deeply embedded in the study of scientific practices, such as Kenneth Schaffner's work on behaviour and *C. elegans* (1998, 2016). Also complementary are views arising from historical and social scientific scholarship such as Richard Burian's important early contribution (1993), Hans-Jörg Rheinberger's long-standing work on the use of whole organisms as units of analysis (e.g., 2010), and Marcel Weber's exploration of *D. melanogaster* (2005). For instance, Weber argues that model organisms should be viewed as a central aspect of science's material culture and are part of a distinctive economy that governs the interactions of scientists who work with them. As he also notes, the features that make certain organisms considered good to use as 'model organisms' go well beyond the material features of the organisms themselves and are highly contingent. Most importantly for our purposes (and his), these contingencies have considerable epistemic implications. In addition, although the initial choice of an organism may have been highly contingent and local, what is critical to its continued use is what Weber calls its 'vindication' (2005, 179), a concept that is closely related to what we explore later in detail in our discussion of attributions of plausibility (6.2).

2.5 What's Special about Model Organisms? Features and Differences from Other Uses of Experimental Organisms as Models

At the core of model organism research are several pragmatic features that warrant closer exploration as they supplement our understanding of the representational power of model organisms. First, the role of evolutionary conservation (particularly genetic but also developmental and mechanistic) is critical to the claims and practices associated with model organisms. For example, highly conserved genomic sequences are those which have been maintained through natural selection and typically go far back in evolutionary time. They thus often relate to the most fundamental biological processes shared by many living entities. As such, a key working assumption in early model organism research was that lower-level organisms with smaller genomes were likely to have highly conserved and more compact forms of the more complex, larger genomes found in higher-level organisms, although evidence was not yet available to provide support for this. Note that such a claim has become more complicated to maintain in light of evidence relating to the C-value paradox: the amount of DNA in a haploid genome (the C value) does not seem to correspond strongly to the complexity of an organism, and C values can be extremely variable. Notwithstanding, such working assumptions were common in the early stages of model organism work, providing justification for their use as models for fundamental biological structures and processes of interest.

It is important to note that few actual relationships between the model organism and the larger group being modelled were recognised or well-articulated in the earliest stages of model organism work, precisely because the detailed genomic sequencing required to analyse the validity of these sorts of claims was yet to occur, although there are some notable exceptions such as homologies in the Hox genes in *D. melanogaster* and other species (see also Weber 2005). Thus, the criteria by which claims of representational scope can be judged to be more or less likely were often external to any particular model organism research project, relying on a promissory note or set of working assumptions about general principles associated with various forms of evolutionary conservation. Notably, choices of model organism also did not hinge on precise knowledge of the phylogenetic placement of a particular organism in relation to others. As various authors have observed, most forcefully Bolker (1995), many of the classic model organisms have proven to be taxonomic outliers (see also Gilbert 2009). However, the zebrafish *D. rerio* arguably may be an exception as it was chosen in part because of its taxonomic placement. In contrast, both the nematode *C. elegans* and the fruit fly *D. melanogaster* have genes that are often very divergent at the sequence level from the homologous genes in mammals for which they are intended to serve as models.

A second defining feature of what makes something a model organism as opposed to any organism that can be used for research purposes relates to characteristics that make doing research with the organism more tractable, as extensively discussed in the historical and sociological literature (reviewed in Ankeny & Leonelli 2018). The usual narrative associated with model organisms is that they were specially selected as research materials because they were viewed as easy and relatively inexpensive to procure, transport, maintain, and manipulate experimentally, especially when compared to higher mammals and primates (which also present more complex, ethical, and affective concerns). The so-called August Krogh principle is perhaps the most commonly cited slogan associated with choice and use of experimental organisms: 'For a large number of problems, there will be some animal of choice, or a few such animals on which it can most conveniently be studied' (Krogh 1929; Krebs 1975; Jørgenson 2001; with reference to model organisms in particular, see Gest 1995). However, this principle tends to be used in diverse and inconsistent ways (Green et al. 2018; Dietrich et al. 2020), and hence its use can obscure the diversity of characteristics present in organisms that are associated with different research programmes.

What is essential in the case of model organisms is that their experimental characteristics are closely related to their power primarily as tools for genetic intervention and manipulation. Again, some exceptions should be noted such as the frog *Xenopus laevis*, which was envisioned as a developmental tool. Model organisms typically have small physical and genomic sizes, short generation times, short life cycles, high fertility rates, and often high mutation rates or high susceptibility to simple techniques for genetic modification. Furthermore, model organisms have been developed using complex processes of inducing particular characteristics in order to establish a standard strain which then serves as the basis for future research. The standard strain, often paradoxically referred to as 'wild type', is a token organism developed through various laboratory techniques (ranging from cross-breeding to genetic manipulation) so that it possesses features valued by researchers and can be reproduced with the least possible variability across generations, for example, through cloning (on such processes in the neurobiology of the nematode *C. elegans*, see Ankeny 2000). Of course, not all of the important biological characteristics of these organisms were evident when they were first obtained in the field (in their truly 'wild' form), but rather they come to be expressed or even induced in the processes of manipulation and standardisation in the laboratory setting. Thus, the ways in which model organisms represent the world is peculiar, if not unique, and strongly grounded in their use as tools for the control and manipulation of biological, particularly genetic, processes.

These standardisation procedures are an essential step in establishing something as a model organism because model organism research hinges on (eventually) developing a detailed *genetic* account of the standard organism in terms of sequence, gene function, phenotype, and so on. This characteristic derived from the historical context in which model organism research was developed and through which the term 'model organism' came to have the epistemic significance now associated with it. Throughout the twentieth and into the twenty-first century, genetics has had a prominent role in biological research and thus has come to define how biologists understand two notions of central importance for developing widely representative models. The first idea relates to what is termed as the 'pure line', which is crucial for the purposes of experimental control over what strains are used, for reducing variability, and for which genetic analysis acts as a defining measurement (Rheinberger & Müller-Wille 2010). The idea of 'comparability' across species has become closely associated to the principle of genetic and other forms of conservation described earlier. A genetically based approach to understanding cross-species comparison and in turn standardisation was not strictly necessary for the conceptualisation of the category of model organisms and their use. However, for reasons that were at least partly contingent, the classical tradition of genetic analysis ended up playing an important role in shaping the experimental practices and concepts used to investigate and standardise organisms (see e.g., Kohler 1994; Weber 2005, 2007 on *D. melanogaster*).

Many experimental organisms do share some of the attributes found in model organisms, particularly those associated with tractability. Some undergo extensive processes of standardisation, and of course biologists may do research on them using genetic methods. However, standardisation is not a defining, generic feature of the broad class of experimental organisms, since how standardised the organism is in genetic or other terms is also a function of the question under investigation. For instance, if one is interested in variations in behaviours of pigeons, the standardisation of specific 'pigeon types' will not be a critical part of developing the experimental organism. By contrast, using frogs for the study of respiration required trying to find organisms with similar morphologies and size, so that their lungs could be studied as though they belonged to the same token animal, and hence involved standardisation processes, albeit not in a genetic sense.

It is clearly not an essential requirement for all experimental organisms to be genetically tractable; again whether this is necessary is a function of what research question is under investigation. For example, some research groups will invest considerable efforts in organisms that are not tractable (genetically or otherwise) according to conventional definitions because they are nonetheless viewed as biologically interesting. For various sorts of experimental organisms, obtaining the organisms on which to do work involves considerable efforts in the field, let alone to grow, maintain, and manipulate them. Researchers continue to use their organisms of choice in part because they think that they are particularly well-suited for the questions of interest: for instance, turtles have characteristics that make them extremely useful for studying transitions from one cell type to another due to the fact that they convert soft tissue into bone (Maher 2009). In summary, the most important criterion for the selection and development of experimental organisms is the way in which they enable the study of specific questions; experimental tractability is also relevant but will be diversely defined depending on the question of interest and is often subsidiary to it.

2.6 Conclusion: Tools for Which Job? Model Organism Research as a Way of Knowing

Our use of the notions of representational scope and representational target broadly parallels the account of models found in the 'models as mediators' account defended by Margaret Morrison and Mary Morgan (1999). The notion of mediation is used to suggest that a model serves 'both as a means to and as a source of knowledge' (Morrison & Morgan 1999, 35): models constitute the meeting point between knowledge and reality, thus providing 'the kind of information that allows us to intervene in the world' (Morrison & Morgan, 1999, 23). In this same sense, experimental organisms are models that mediate between theory and the world. The theory or question to be investigated is the representational target, and the 'world' that the model represents can be defined in terms of its representational scope. Such scope may be quite delimited, for instance, to understanding the phenomenon in question within a certain group such as mammals, or much broader, as is the case with model organisms.

Representational scope and target can vary not only organism to organism, but they also over time with regard to a specific organism during the process of research. Indeed, Rheinberger (1997) and others (2000; Morgan 2003, 2007) have pointed to the ability to lead researchers in unexpected directions as one of the main attractions of working on real organisms in the lab. Experimental organisms have been engineered and modified to enable the controlled investigation of specific phenomena, yet at the same time they remain largely mysterious products of millennia of evolution, whose behaviours, structures, and physiology are often still relatively ill-understood by scientists. Through this hybrid status as both natural and artificial objects, experimental organisms facilitate exploratory research by enabling biologists to ask questions without necessarily having clear expectations about what answer they will obtain or even about what questions will end up being the focus of inquiry (on the theoretical issues associated with this type of 'exploratory experimentation', see Burian 1997; O'Malley 2007).

Model organisms are an important subset of experimental organisms with very particular qualities and representational power, which include a mixture of features intrinsic to the organisms themselves, features derived through the manipulation of organisms for research purposes, and features attributed to organisms by the researchers who use them. We summarise the characteristics contributing to the establishment of a model organisms in Table 1. Rather than being generic tools for experimental interventions, model organisms in fact represent a unique 'way of knowing', in John Pickstone's terms (2001). They involve a set of essential commitments, features, and practices that emerged in relation to a set of distinctive epistemic goals. These in turn have been finely tuned to the study of the objects that these models are taken to represent, namely, shared fundamental biological phenomena. Model organisms are the right tools for a very specific type of scientific job, that of investigating and manipulating organisms that are kept in isolation from their natural environments. What is critical to understanding their unique status is that they are grounded in epistemic commitment to pursuing integrative and comparative accounts of life by focusing on individual organisms as the main unit of analysis.

3 How Do Model Organisms Represent?

3.1 Introduction

This section explores how generalisable arguments are made through abstracting from individual specimens recognised as model organisms and provides an overarching framework for understanding how experimental interventions on these organisms inform the development of biological theories and the scientific understanding of various life forms. We explore the activities associated with abstracting, including the ways in which theory informs (but does not

Table 1. Characteristics contributing to the establishment of a model organism

Characteristics of the Organism	*Natural or intrinsic*	Tractability in the lab Length of life cycle Fertility rates and ease of breeding Size of organism Ease of storage Size of genome Physical accessibility of features of interest
	Induced/uncovered through experimental interaction and transfer to lab	Mutability of specimens Response to lab environment (food, light, temperature, cages, routine) Availability of standardised strains
	Attributed to or projected onto the organism by researchers	Representational scope (how extensively the results of research with the organism can be projected onto a wider group of organisms) Representational target (number of phenomena that can be explored via the organism) Power as genetic tools Ability to serve as the basis for comparisons to other organisms Multi-disciplinary usefulness (capacity to fit different research domains, e.g., genomics, development, and physiology) leading to cross-level integration

determine) these activities. We argue that model organisms play an anchoring role that arises out of their dual status as both samples and artefacts, particularly because of their highly controlled variability that creates considerable limitations on their relationships to the wider environment. We also explore the ways in which model organism research is comparative in a very particular sense, namely, in how it sets boundaries on comparisons to make them more product- ive through exploiting variability in its narrowest sense. Model organisms thus are transformed into models within highly standardised, uniform, and simplified environments, which because of their 'placelessness' can function as anchors for a broad and ever-evolving modelling ecosystem. The representational power attributed to model organisms hence shapes the research practices within which they are used. Related conceptual commitments become entrenched in the ways in which biologists theorise and perform material interventions in the world, such as genetic and phenotypic manipulations. These factors in combination make model organisms into potential models for a very wide variety of phenomena.

3.2 Making Organisms into Models

Model organisms as research tools have an ambivalent status; they are simultan- eously artefacts and samples of nature. On the one hand, specimens of model organisms are actual organisms: they are entities that we could not hope to create from scratch in a laboratory (despite many attempts to do so via robotics and synthetic biology techniques), precisely because we understand only a minimal part of how they work in most cases. Model organisms have the power to generate surprising results, both in terms of their representational target (as when signal- ling pathways in the zebrafish *D. rerio* turned out to be useful to study the onset of Alzheimer's disease) and representational scope (e.g., the 1983 discovery that certain *D. melanogaster* sequences, such as homeobox, are conserved not only in fruit flies but also across the animal kingdom). As such, model organisms are favoured materials for exploratory experimentation. They remain samples of the very part of nature that they are taken to represent, that is, they are samples of the variability present in a natural population.

On the other hand, the transition of any organism into a research environment is accompanied by a series of modifications to the organism itself, particularly in cases where researchers plan to use the organism in the long term and over many generations, and when they have specific representational demands (as detailed in the previous section). The process of preparation of an organism for experimental use requires the selection of traits on which researchers wish to focus (and thus to stabilise and control), such as the zebrafish *D. rerio*'s transparent skin and ability to absorb drugs poured into aquarium water. This process starts from the selection of specimens as research materials. The individuals chosen to populate a lab need to conform, at least in some respects, to the expectations of researchers that intend to experiment on them. They must display features that are appropriate in combination with the research procedures and instruments in use. They also need to adapt to the climate of the storage facilities where they are kept, which are often geared towards generating standard reactions (e.g., in the case of plants, regulating their circadian rhythms via the lab's lighting conditions).

Specimens initially adopted by researchers as model organisms never conform to all of these expectations, and are typically manipulated in a variety of ways ranging from genetic to environmental interventions, so as to fit these requirements as well as possible. Standards of care and maintenance also need to be developed for specimens to retain these characteristics through generations, and thus remain relatively stable biological platforms for biological investigations (Rosenthal & Ashburner 2002; Leonelli 2007b). It is through these diverse activities that the traits displayed by a few individual organisms become models not just for their own taxon, but for different kingdoms. Obtaining specimens that conform to researchers' expectations thus requires relevant techniques, standardised tools and guidelines, and extensive experience in handling the organisms: it is a matter of skilful production, rather than mere convention, transfer, and use.

The dual status of model organisms – at once samples of nature and human artefacts, simultaneously modelling known and unknown phenomena – is the feature that makes them such interesting objects in biological research, and indeed an important and distinctive type of scientific model. They are highly domesticated samples of nature, whose handling and traits become so familiar to the researchers employing them as to become 'tame'. Organisms are reproduced and modified under such controlled and purpose-oriented conditions that they may end up bearing relatively little resemblance to their relatives in the wild: their features have been largely reshaped by scientists according to their research needs, and yet they include processes and entities that are yet to be understood by researchers.

Manipulation is crucial and strongly underpins representational choices and directions (see also Love & Trevisano 2013). The material modification of the physical features of organisms to create tractable and representative models can be understood as involving processes associated with abstracting. In other words, such modification involves the transformation of some features of a phenomenon into parameters used to model it, depending on the specific aspect of biology that the model is deployed to study. For instance, a trait such as 'short life cycle' or 'experimental tractability' is identified and treated as a parameter for the selection and use of a model. Thus, the model is developed to instantiate that trait in the clearest and most effective way possible, for example, by eliminating strains that exhibit longer life cycles and higher levels of vulnerability to life in the lab. Defined in this way, abstracting is one of the processes required to create a model, rather than an attribute of the model itself. In other words, the model is 'abstracted' in various ways depending on the specific circumstances and research goals in any particular case rather than being 'abstract' in an absolute sense (Leonelli 2008). Further, abstracting is essential in the context of modelling practices, as it is the process by which any material model acquires representational value with respect to some aspects of a phenomenon.

Maintaining control over the development of traits characterising different individuals ensures the replicability of specimens with particular traits, as well as the stability of their features regardless of the time and location of their use. Abstracting involves physical interactions between the researchers and the objects to be modelled, including selecting a limited set of material features of organisms as potentially interesting for research purposes. It involves devising ways in which these properties can be incorporated into a unique specimen, making certain that specimens with those characteristics can actually be stored and safely kept in the available research space, and constructing a toolkit of guidelines, materials, and instruments that allow researchers worldwide to obtain and maintain uniform specimens. These conditions are realised in part by modifying the environment of the organism, including living spaces, nutrition, light, and other husbandry conditions. For instance, *A. thaliana* ecotypes are expected to have uniform height and developmental schedules, which are generated by providing the same growth conditions for all plants and isolating strains to prevent cross-breeding. Direct interventions on the organisms themselves, such as genetic modification, also are involved. Researchers may eliminate plants with unexpected leaf shapes to control for the risk of unexpected mutations in the population to be studied or use bacteria to generate more mutants with surprising traits. In this manner, what is eliminated and abstracted away is some of the population-level variability.

Background theoretical knowledge is involved in researchers' choices of which traits to abstract and reproduce in the models: abstracting is clearly theory-informed (Waters 2007). The theoretical commitments made while developing material models such as model organisms thereby become entrenched in the subsequent uses of these models as laboratory tools, sometimes with significant implications (see Section 7). However, abstracting is not theory-guided: theoretical knowledge does not wholly determine the activities and results of modelling. The manipulation of models and the selection of traits to be modelled require only some interest in exploring one or more aspects of the phenomena that they are taken to represent, and the processes tend to be highly descriptive (Ankeny 2000). Specimens are taken to be representative of a set of phenomena dependent on the research context. Epistemic access to phenomena is granted first and foremost by material manipulation, since the amount of intellectual manipulation necessary to handle these models is minimal. Material models obtained in this manner thus constrain and enable investigation as well as the formulation of research questions, modes of intervention, and choices and use of instruments and methods.

3.3 Controlling (Not Eliminating) Variability: Model Organisms as Families of Specimens

An immediate question arising from the abstracting of individual specimens into model organisms concerns the significance of variation not only among individual specimens but also among different strains of the same species. While most model organism researchers (particularly those active between the 1980s and early 2000s) have focused on few highly standardised strains of the same species, and the strain used to produce reference genomes typically becomes the most popular, model organism work rarely involves focus on only one strain. Rather, it typically involves comparisons across results obtained through experimentation on two or more strains of the same organism, which help to identify significant mutations and to assess whether (or not) they may be conserved. For example, much *D. melanogaster* work happens on two particular strains (Canton-S and Oregon-R) but in parallel with other strains; the Columbia ecotype of *A. thaliana*, used as reference for its sequencing, is flanked by the Wassilewskjia and the Landsberg *erecta*, which are also popular with researchers. In addition to this heterogeneity, there is also the facility with which the original specimens that have been abstracted and standardised for experimental purposes themselves acquire variants: again in *A. thaliana*, we find several variations on the Columbia ecotype that all derive from the same lineage but are likely to have maintained different polymorphisms and accumulated different mutations.

This variability should not be surprising, given that researchers are working with a 'live' model, an entity that develops and evolves, and whose dynamic, processual nature continuously defies reification (see Nicholson & Dupré 2018). Indeed, taking an organism as a model does not involve transforming it into an inanimate object with fixed characteristics: stabilising some of its features is necessary to focus on the types of variation of interest to researchers. In this sense, model organisms are best understood as indicating a family of material objects with very similar characteristics and a common phylogeny. These families of objects provide and delimit a space for comparison, while also functioning as a (presumed) stable material platform for experimentation across different sites and times.

A critical implication of this point is that model organism research does not exclude comparative approaches, as sometimes claimed in the literature. Rather, model organism research strongly delimits comparison in order to fit the very narrow variability fostered and admitted within the models being produced via abstracting and standardisation: it sets boundaries on comparisons to make them more productive. This approach relies heavily on the commitment to evolutionary conservation that is central to research with model organisms. Thus, comparative analysis exploits variability in its narrowest sense, and largely from the point of view of experimentalists interested at least in part in molecular approaches. At the same time, the enduring significance of comparison – even in this highly delimited form – cuts across rigid distinctions between experimental and naturalistic approaches. Model organism biology clearly relies on the use of few species to make claims of wider validity (what historians Bruno Strasser and Soraya de Chadarevian (2011) call the 'exemplary' method), yet also appeals to the legacy of natural history in the ways in which it capitalises on comparison across closely related cases (see also Section 7.2).

In discussing the comparative methods at work within recent evolutionary morphology, James Griesemer (2013) has highlighted the significance of model taxa as 'material platforms for a research system on which to conduct integrative science' (526). We claim that model organisms function in much the same way, but that the characteristic narrowing of variability and the reliance on genetically grounded comparison associated with model organisms enables researchers to do the kind of multi-level integration that became fundamental to developing interrelations between molecular, developmental, physiological, ecological, and behavioural approaches at the start of the twenty-first century.

3.4 A Modelling Ecosystem: How Model Organisms Facilitate Integrative Understanding

Recognising the diversity of objects that can be encompassed by the idea of a 'model organism' does not undermine our abilities to understand model organisms as scientific models. Rather, recognition of this multiplicity helps us to highlight something that many philosophers have observed in relation to biological research: one model is never enough. Modelling strategies in biology are extremely varied in both the form that they take and the ways in which they create bridges between theories and data (Griesemer 1990; Leonelli 2007a; Green 2013). Most research projects require the employment of several types of models, as well as several models of the same type, to achieve their goals.

Model organisms are part of a much larger modelling ecosystem, and the manipulation of material organisms anchors a multiplicity of modelling practices that include mathematical, theoretical, and diagrammatic models (see, for instance, Meunier 2012 for the case of zebrafish *D. rerio*).

Why do model organisms play this anchoring role? Their ability to play this role arises from their dual status as both samples and artefacts, which makes them into potential models for a wide variety of phenomena. Indeed, it makes it possible for researchers to attribute a wide representational scope to them. There is no straightforward pairing between model organisms as material models, and any one model description (as required, for instance, in Weisberg's 2013 framework); the opportunity to directly intervene on organisms as material models further secures their epistemic value as integrative platforms. These models encompass countless aspects of the world (potential phenomena) that come under scrutiny for different purposes by various types of biologists.

How does this process work? It functions through limiting variability dramatically not only in terms of the variability directly associated with these organisms but also with their relationships to their wider environments. In fact, the standardisation of the model organisms' environments – the fact that exposure to natural changes in climate, nutrients, lighting, and other factors is typically limited if not altogether eliminated for model organisms – constitutes the biggest source of uniformity. It also has crucial conceptual aspects: model organisms are not models of organisms situated in their natural environments; instead model organisms are *separated from* their natural environments. What researchers want to study is how organisms develop under standard laboratory conditions. The variability originating from relations to other organisms (including via the microbiome), soil, climate, and so on is eliminated. From their origins as organisms that are co-dependent on a broad and complex biological ecosystem, model organisms are transformed into models within highly standardised, uniform, and simplified environments, which because of their 'placelessness' can function as an anchor for a broad and ever-evolving modelling ecosystem. What is important here is the relation of model organisms with the research environments in which they are being studied and manipulated, which includes other models as well as the laboratory conditions, methods, tools, and infrastructures that come to constitute their new milieu.

The development of infrastructures, techniques, and mathematical models specific to model organisms that facilitate their study is historically and epistemically intertwined with the development of the actual organisms as material models with specific physical characteristics. The choice of which properties of the original specimens should be abstracted and retained into the standardised model was made partly on the basis of researchers' theoretical interests and partly on what available methods and infrastructures made possible, which in turn has changed through time (Leonelli & Ankeny 2012). As the techniques, lab conditions, and instruments built to interact with and use model organisms have become more sophisticated and specialised, researchers increased their abilities to control the organisms themselves and in turn their abilities to standardise and stabilise specific characters.

Repeated use of and reference to similar organisms kept under the same environmental conditions provides considerable opportunities for sharing knowledge (including know-how) across a vast constellation of biological disciplines, groups, and research schools. This sharing of knowledge is achieved through formal means such as peer-reviewed publications, but importantly also through more informal processes of communication such as grey literature and lab visits aimed at learning skills and techniques. The research environments within which these instruments, data, methods, and infrastructures are used can be assumed to be the same or at least to be broadly reproducible. All of these factors make model organisms into movable resources that can be easily resituated (Kohler 1994). Through this shift from a biological to a modelling ecosystem, model organisms become low-cost, low-maintenance research materials that are easy to control and on which a substantial body of knowledge can rapidly be accumulated.

This type of abstracting is what makes model organisms into platforms through which several other types of models can be related and integrated. Coordination can be achieved via different domains of questions asked by the biologists involved as well as the acquisition of common epistemic skills used in laboratory work, thereby providing a strategy to navigate diverse theoretical and methodological commitments, integrate know-how with theoretical knowledge (Leonelli 2009), and pursue common goals (Love 2008). It can therefore facilitate the acquisition of integrative understanding and underpin interdisciplinary collaboration across domains as diverse as molecular biology, physiology, development, and even ecology (Bevan & Walsh 2004). Model organism research is now a canonical example of multi-level research, which includes the ability to relate multiple conceptual, methodological, and explanatory perspectives to one another (Mitchell 2003) as well as the integration of causal-mechanistic and mathematical models representing findings pertaining to different levels of organisation of the organism, ranging from the molecular to the cellular and developmental (O'Malley et al. 2014).

3.5 Model Organisms as Models

In the previous section, we emphasised that model organisms function as representations of both other organisms (their representational scope) and the organism taken as a whole (their representational target), and this duality lies at the core of their representational power as models. In this section, we have considered the characteristics of the material objects that constitute model organisms and the ways in which they relate to a broader modelling ecosystem and research goals. Let us now bring these elements together to further clarify how model organisms function as models of life. To do so, we build on Roman Frigg and James Nguyen's Denotation, Exemplification, Key and Imputation (DEKI) model of representation (2018) and provide a more formal character-isation of how model organisms, as models, represent both other organisms and the whole organism, which we summarise in Figure 1. We then defend this representational role as foundational to the functioning of model organisms as *models*, including the many ways in which model organisms are used as *tools* for the study of other organisms. We thus propose to view the role of these models as representations and as research tools as one and the same: model organisms are not useful in biological practice without an underlying commit-ment to a specific form of representational power. Such a commitment unavoid-ably affects the ways in which these models are employed to study and manipulate biological processes.

A family of individual specimens, typically displaying similar phenotypic properties including their appearance, genetic make-up, and growth/develop-ment patterns, constitutes the material object that functions as a model. What makes this object into a model is an underlying commitment to the idea that a combination of properties of the object (e.g., colour, metabolism, circadian

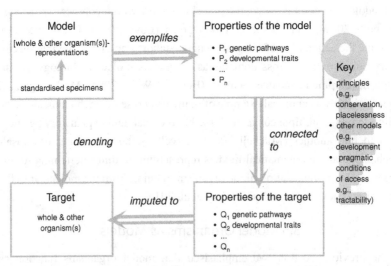

Figure 1 How model organisms represent (adapted from Frigg and Nguyen 2018, design by Michel Durinx)

rhythms, and genetic traits) represent the properties of the assumed target (other species and the organism taken as a whole). The object is thus interpreted as a type of representation for the intended target: in our case, the model consists of the positing of the object 'standardised specimens' as a 'whole and other organism(s)'-like representation. The commitment to the idea that properties of the object represent the properties of the assumed target is exemplified by specific properties of the model, which include, for example, conserved genetic pathways, complex developmental mechanisms, and specific causal relations between a given gene cluster and phenotypic traits. In the course of their work, researchers impute these properties to the target of their study (the biological phenomena in which they are ultimately interested), thus solidifying the representational relationship between model and target.

What makes this crucial passage possible is commitment to a 'key' that allows researchers to connect properties of the model (P in Figure 1) with properties of the target (Q in Figure 1). The key specifies why, how, and under which conditions the properties of the model that have been singled out by researchers, such as conserved developmental pathways, can legitimately be attributed to the target. The key typically associated with model organisms typically includes several factors such as principles (e.g., evolutionary conservation); the fit with other models such as simulations, diagrams, and mathematical models of development; and pragmatic factors such as the extent to which the objects chosen as models make the properties tractable and accessible. Frigg and Nguyen view researchers who use models as free to choose whichever key they may like and find useful. In the case of model organisms, the key emerges from consideration of the physical (and partially abstracted) features of the model as well as commitments and habits adopted by model organism communities over the course of decades. It thus makes choices more social than individual (for a study of the processes of scaffolding and entrenchment involved in such cultural developments, see Caporael, Griesemer, & Wimsatt 2014). As we argue in Sections 4 and 5, individuals need to accept the key for their work with model organisms to be recognised and sanctioned by their peers: we will come back to the significance of this observation later in the Element.

Another important characteristic of this approach to modelling is its agnosticism towards the ontological status of the target. We prefer to construe the target of the model (the phenomena that the model is taken to represent) not as a literal description or embodiment of the world, but rather as the result of researchers' efforts to conceptualise their interactions with the world. Thus in our account the target 'whole organism' refers to any one organism in the world, but it does so through a theoretical perspective that is embedded in the language used. This interpretation provides one way to understand our framework; it is perfectly compatible with regarding targets as actual parts of the world or alternatively as literal, truthful descriptions of the world.

Frigg and Nguyen note that the fact that 'a model as a whole denotes a target as a whole does not preclude there being additional denotation relationships between parts of the model and parts of the target' (2018, 14). We agree with this assessment in the case of model organisms. The model as a whole – the family of specimens under investigation by the researchers – represents the 'whole organism' and 'other organisms', which is perfectly compatible with specific features of the model (e.g., the ways in which cells divide in a given yeast strain) being themselves used as models for specific features of the target (e.g., the proliferation of cancer cells in humans, or the mechanisms underpinning patterning in embryo development, e.g., Meunier 2012). These specialised models are part of larger modelling ecosystems that model organisms help to anchor and integrate. Most model organism research does in fact focus on such specialised models, as researchers focus on one selected subgroup of questions (and part of the organism) at a time. The commitment to using model organism specimens as models facilitates the material realisation of these specialised projects, since it provides researchers with the necessary background knowledge and appropriate key to link properties of the target with properties of the model. It also makes it possible to integrate the results of specialised projects into a broader integrative understanding of the organism as a whole, and the ensemble of techniques and methods of biological intervention developed through the experimental manipulation of these models.

The DEKI schema is particularly helpful in the case of organisms used as models because the object that forms the base of the model (the particular organisms in question and their properties $P_1, P_2, ...$) is easy to confuse with the phenomenon that is being investigated (which is not the organisms themselves, but rather specific clusters of properties attributed to a wide range of organisms and to organisms taken as wholes). Being specific about what elements are involved in any particular instance of denotation, and the fact that the object used as a representation is not the same thing epistemically as the phenomenon being represented (even when these two things may appear to be the same in practice), is crucial not just philosophically but in terms of the precise types of claims being made about model organisms by researchers. Thus, this account undermines oversimplified notions of model organisms acting as 'general models' or as straightforward embodiments of phenomena without explanation or interpretation. It instead emphasises both the extent to which the use of organisms as models is theory-informed, and the extent to which it can involve creativity and novelty in terms of the types of knowledge and insights obtained.

Furthermore, this account of modelling provides a way to situate representational claims in relation to experimental practices and know-how. Recent philosophical scholarship on modelling has (rightly) moved away from treating representation as the main goal and measure of excellence for research (e.g., Cartwright 1989; Woody 2000; Knuuttila 2011). Our analysis of the significance of experimental practices and cultural understandings of model organisms underscores that we agree with this emphasis on the *use(s)* of models as paramount in determining their epistemic roles and status. Clearly, there are many uses of model organisms that are not tied solely or primarily to their representational power. For instance, homeobox genes from *D. melanogaster* are not representations of partially homologous genes from other organisms, nor are they used as representations: they are important tools for identifying more genes that might play similar roles. They have led to the discovery of homeobox-containing genes in hundreds of other metazoans, but also often end up failing to represent because such genes are not found in the organisms of interest.

Similarly, it could be claimed that probes or methods resulting from model organism research, such as RNA hybridisation probes, gene mapping techniques, or knockout experiments, are not tied to the representational power of model organisms as models, but rather to their versatility as laboratory tools.[4] However, even in these cases, we argue that the representational power attributed to model organisms continues to define the epistemic significance and implications of adopting such organisms as models. Building on DEKI, we are interpreting the representational power of the model as itself grounded in their use as research tools and the related habits and commitments made by model organism communities. Thus, while it is true that homeobox-containing genes in *D. melanogaster* do not always serve as reliable representations of the genetic make-up of other organisms, the use of *D. melanogaster* itself as a model organism is the reason why such genes are sought in other organisms, and why they can in fact serve as tools for identifying additional genes that might play similar roles. The adoption of *D. melanogaster* does, in turn, carry specific conceptual commitments, such as the attention to genetic mechanisms over and above the susceptibility of the organism to environmental changes, resulting from the representational power of this model. The same applies to the adoption of probes and techniques derived from model organism research: these carry with them specific commitments to how model organisms represent a burden

---

[4] We are indebted to an anonymous reviewer for articulating this important objection, which enabled us to clarify this crucial aspect of our account.

whose significance for future research is often overlooked by researchers, and yet must be critically assessed.

We conclude that understanding how model organisms represent is crucial to understanding their role as tools to study other organisms. As Ian Hacking put it nearly four decades ago (1983), modes of representing and intervening are tightly interconnected in scientific practice; the case of model organisms elegantly instantiates this insight. Model organisms are material models with a specific representational power, which facilitates and underpins their use in biological practice. This link between the representational and interventionist dimensions of the model is crucial to its power and its potentially pernicious effects on research. Whenever researchers use model organisms, and/or techniques, probes, and infrastructures produced in association to those models (i.e., the repertoire that we shall analyse in Section 5), they commit to the conceptualisation of organisms that we outlined in the previous section; in other words, they commit to a view of organisms as genetically conserved, placeless, and highly standardised. This commitment affects the subsequent study and manipulation of biological processes, for instance by making it harder to fit results coming from model organism research into studies of biological variability in relation to changes in climate, ecosystem, and soil composition (as we shall see in more detail in Section 7).

At the same time, formalising our analysis in this way provides no insights on whether model organisms are in fact *good* models for their biological targets, or more precisely, how researchers determine whether a model is likely to be correctly denoting its target and what grounds they use to make these sorts of judgements. As Frigg and Nguyen explicitly acknowledge (2018, 14), what establishes that a model denotes a target is not easily explained as part of their framework. In our view, answering this question requires complementing their abstract, a temporal schema with a study of the actual practices of model organism research as a way of knowing, which we do in Section 4. A crucial epistemic consideration emerging from considering practices in detail is the iterative nature of the relations of representation expressed in Figure 1, and their implications for experimental practice. As we show, commitments made around what constitutes the model (in a material sense and in the sense of what it is taken to represent), what constitutes the target, what properties the model is taken to exemplify, and whether and how those properties are imputed to a target can change. These commitments result from the conceptual, material, and social developments within the research communities in question, many of which become entrenched over time and strongly affect experimental practice. We return to the question of what makes a 'good' or plausible model in Section 6, providing an updated version of Figure 1 that takes these aspects into account.

3.6 Conclusion: Modelling Life

Infrastructures, multiple models of various kinds, and model organisms are co-constructed as part of the same way of doing research and as part of the same push towards standardisation. This mode of investigation becomes entrenched, generating families of models that work with each other and can be easily adopted as ways of doing biological research (Wimsatt 2000). Thus, model organisms by themselves do not provide an integrative, holistic, multi-level understanding of biology; rather, it is the modelling ecosystem that makes it possible for model organisms to function as reference points for biological integration across levels. In turn, the development and upkeep of this modelling ecosystem depends on the resourcing, organisation, and institutionalisation of research communities and practices, which we discuss in Section 4. The use of model organisms is a peculiar and oftentimes controversial way of modelling and understanding life. Examining who developed, adopted, and accepted this way of knowing is crucial to assessing its advantages and limitations, as well as the grounds on which model organisms can be regarded as plausible models in the first place, which we examine in Section 5.

4 For Whom Do Model Organisms Represent?

4.1 Introduction

The history of each model organism differs in several respects, not least due to the variation in habits and institutions characterising the researchers who work on different kingdoms (plants as compared to animals or fungi, for example). Despite this diversity, distinct 'model organism communities' have formed around particular organisms in order to profit from the expertise, methods, instrumentation, and data accumulated by participating biologists. These communities differ in size, internal structures, and degree of formality, among other factors, and yet they share a set of norms and attributes that are closely connected to the epistemic value associated with model organisms. In this section, we explore the common norms and institutional mechanisms fostered within model organism communities to encourage collaboration and the sharing of resources. We discuss the ways in which these norms and practices have been circulated and refined through discussion both within and across these communities, and the epistemic implications of these developments.

4.2 A Common Vision and Ethos

The biologists credited with beginning the use of a specific organism as a model organism typically have been charismatic leaders with strong scientific skills, whose influence on the subsequent development of the community was so strong as to warrant the label 'founder effect'. These biologists also possessed sophisticated political and organisational abilities which helped to attract considerable support from peers, funders, and institutions. In some model organism communities, attention tended to focus initially on a particular individual: think of Sydney Brenner for the nematode (*C. elegans*), George Streisinger for zebrafish (*D. rerio*), and Paul Nurse for fission yeast (*Schizosaccharomyces pombe*). In other communities, a cast of energetic founders came into play, as in the case of *A. thaliana* with Frederick Laibach and George Rédei, where early efforts were continued and dramatically expanded in a highly coordinated fashion by a handful of highly influential researchers based in the United States and Europe between the 1960s and the 1980s (see Leonelli 2007b).

These individuals shared keen interests in exploiting the opportunities presented by molecular approaches to investigate broader questions around the biology of organisms including developmental and evolutionary dimensions. They sought to develop detailed understandings of their respective organisms in order to use results and insights for application in other domains, either to more complex organisms including humans or to make generalised claims about biological processes and phenomena. Such approaches were thought to be appropriate and likely to be effective due to shared (but largely untested) assumptions about evolutionary conservation (see Section 2.4). These researchers also shared a strategy to investigate organisms in a complex, interdisciplinary way: they advocated a holistic, inter-level approach to organisms in isolation from their environment (which as discussed in Section 3 meant paying little or no regard to organisms' living conditions, as long as they were stabilised and kept uniform within any one laboratory environment). This integrative vision was crucial to shaping the conceptual goals associated with the model organism way of knowing. These goals have shaped the ways in which model organisms are not only viewed, but also how they are used, as models. That is, these epistemic goals are closely tied to the representational scope and target of model organisms, as discussed in Section 2.

The founders of model organism communities also shared a strong sense of the norms that should govern scientific communications and interactions. The actual implementation of such norms differed from community to community both in terms of their extent and relative success, as has been adequately documented in historical scholarship. They tended to emphasise mechanisms and processes to support efficient and collaborative divisions of labour among groups focused on the study of the same organism. These sometimes included relatively controversial practices such as sharing data and insights at the pre-publication stage, engaging with a wide international network when planning research to ensure that research efforts would not be duplicated, and generally favouring collaborative over competitive behaviours. This ethos or moral economy (Kohler 1994) was aptly summarised as 'share and survive' (Rhee 2004), and explicitly opposed to the 'publish or perish' motto that underlay much animal-based research in the 1980s and 1990s.

It is striking that the ethos of sharing, now widely recognised as a distinct characteristic of contemporary model organism communities, was present even in the absence of digital communication technologies. Until the early 1990s, coordination within each model organism community occurred among what were relatively small groups of researchers through newsletters, meetings, lab visits, personal contacts, and so on. For instance, from the 1980s onwards, researchers working on the nematode *C. elegans* used a newsletter to distribute information on techniques and strains as well as pre-publication results (on similar mechanisms in *D. melanogaster*, see Kelty 2012). They also published handbooks on worm biology authored by 'the Worm Community' (Wood et al. 1988) to capture all work to date and to underscore their communal approach to research. In addition, in early *C. elegans* and related nematode research, publications were delayed because many projects were very large in scale and took considerable time, and the funding structures associated with the main UK lab did not require prompt journal publications (de Chadarevian 2002; Jones, Ankeny, & Cook-Deegan 2018). Thus more informal sharing mechanisms helped to maintain the coherence of what became a geographically diffuse yet intellectually unified group, in part to avoid costly research duplications but also to permit free exchange of important findings. The founders of the *A. thaliana* community, particularly Rédei and Chris and Shauna Sommerville, enforced the sharing of results at the pre-publication stage of research from the very start of molecular work on the plant in the 1970s (Leonelli 2007b). These efforts laid the groundwork for more sophisticated methods of sharing that arose in the 1990s–2000s.

4.3 Resourcing: Fitting Into the Political Economy of Biology

Most popular model organisms have enjoyed relatively steady governmental funding, particularly during the HGP that resulted in the formal recognition of a set of these organisms as 'model organisms' and enabled many of the associated communities to flourish and expand exponentially. Whether funding for model organisms has resulted in reduced support for research with other organisms or even particular types of biology has been a matter for considerable debate, with results depending on the particular metrics utilised (we do not engage with this literature here, but for discussion see e.g., Davies 2007;

Dietrich, Chen, & Ankeny 2014). The period from 1980s to the 2000s thus has been dubbed the 'age of model organisms' by biologists themselves (Davis 2004).

The proponents of the key model organisms were able to convince colleagues, peers, and funders that repeated use of and reference to the same organism provided critical opportunities for sharing knowledge, materials, and technologies across biological disciplines and research groups. It allowed growth of comparative research, and indeed constituted an anchor around which entire research communities could be built (Ankeny & Leonelli 2011). The leaders of these communities also had significant public relations skills which allowed them to sell their projects well beyond the scientific community that allowed funding and attention including from a large amount from governmental and smaller amounts from non-governmental entities including foundations and industry. In addition, funding was in many cases global; the international HGP started a sort of 'arms race' between different countries with competing prominent labs eager to participate in this novel big science effort. Norms from model organism work being promoted more generally through human genomic projects was in part due to leadership from those in the original *C. elegans* community (Jones, Ankeny, & Cook-Deegan 2018).

In the case of *A. thaliana*, the strong professional and personal bonds established among its advocates enhanced the profile of the plant. It became possible to devise and successfully implement common strategies to obtain funding from national and international bodies (such as the U.S. National Science Foundation and the European Commission), thus implementing the 'share and survive' ethos discussed in Section 4.2. This model of research fostered major infrastructural resourcing such as databases (e.g., The Arabidopsis Information Resource [TAIR] created in 2000 based on previous, less systematic efforts) and steering committees (the Multinational Arabidopsis Steering Committee [MASC]). These were jointly devised by representatives of each country active in plant biology in order to coordinate *A. thaliana* research projects around the world.

These efforts also fit well with even broader socio-technical and political–economic regimes, institutional structures, and moral economies governing the uptake and financing of science and technology during this period, and the vision of basic molecular research as critical for innovation in medicine and agriculture (Cook-Deegan 1994; Hilgartner 2017). Parallels were drawn between the HGP and other big-scale science initiatives such as the Manhattan Project; advocates noted the need for investment in basic science such as genomic sequencing as had occurred with fundamental physics. Critics highlighted fears about top-down, centralised funding, speculating that these efforts created focus on high-profile national goals in part to distract from other problems, such as the failed 'war on cancer' and contemporaneous fears relating to the emergence of HIV/AIDS (Rosenberg 1996).

On the ground, the importance of institutional buy-in cannot be underestimated; numerous prominent institutions were willing to invest in model organism research through hires, capital investment, infrastructure support, and so on. In the case of *C. elegans*, these efforts began with Medical Research Council funding at the University of Cambridge from the 1960s onwards, with multiple major labs subsequently supported in the United Kingdom, the United States, and beyond. They were followed by the creation of what became the Wellcome Genome Campus, including the Sanger Sequencing Centre just outside of Cambridge (de Chadarevian 2002). These jointly oversaw the HGP along with hosting several model organism projects. The German, French, and Japanese arms of the HGP invested heavily in non-human organisms: by 1997, a substantive of their funding was dedicated to subprogrammes focused on mouse *M. musculus*, rat *R. norvegicus*, fruit fly *D. melanogaster*, and zebrafish *D. rerio*, with similar programmes arising elsewhere.

4.4 Digital Infrastructures: Databases

The advent of the HGP in the 1990s, and the availability of funding for sequencing projects on what became the canonical model organisms attached to it, played two important functions with respect to the collaborative strategies of model organism communities. First, they consolidated and institutionalised their epistemic goals. The pursuit of sequencing data was a powerful shared aim, as well as a common denominator that could serve as the basis for future collaborative work. Biologists involved in these projects agreed that gaining access to sequence data was of central importance for future research in all areas of biology, thus constituting a collaborative platform for the integration of knowledge about single organisms as well as for comparative research across species. Second, the new wave of funding for sequencing also facilitated the use of new digital technologies to support the sharing ethos of model organism communities even as they expanded well beyond the initial informal and relatively contained networks to become very large groupings. As conferences organised around research on specific organisms became large (numbering thousands of delegates in the cases of *A. thaliana*, *C. elegans*, and *D. melanogaster)* and publication numbers relating to each organism ballooned (see data in Dietrich, Chen, & Ankeny 2014), researchers turned to digital infrastructures such as databases and repositories in order to facilitate immediate data sharing on a large and efficient scale.

Through application and dissemination of the ethos of sharing, the pre-existing networks associated with what became model organism communities helped to shape what became the community databases associated with many of the model organisms in the late 1990s and early 2000s (Leonelli & Ankeny 2012). Their immediate goal was to store and disseminate genomic data, which became more formalised in the context of the 1998 Bermuda Principles mandating daily release of HGP-funded DNA sequences into the public domain (Jones, Ankeny, & Cook-Deegan 2018). The longer-term vision associated with these databases reflected the deeper goals related to the ethos of these communities. They aimed to establish agreed terminology and language; to incorporate and integrate all data available on the biology of the organism in question within a unique dataset, including data on physiology, metabolism, and even morphology; to promote cooperation across communities and databases so that the available datasets eventually would be comparable across species; and to gather information about laboratories working on each organism and the associated experimental protocols, materials, and instruments, thus providing a platform for continued community building as they began to grow rapidly in size and dispersal.

The focus on sequencing data also presented some additional issues: such data taken on their own could not provide meaningful functional information about the biology of organisms, and hence highlighted the limitations of the typical credit attribution systems based on academic publications which typically require communication of claims and hypotheses attached to specific biological phenomena (Hilgartner 1995; Leonelli 2010). Reconsideration of what makes a 'publishable unit' was a critical part of the transition in practices that occurred during the era of large-scale genomic sequencing, particularly with model organisms, as well as changes in how authorship was determined to allow for extremely large groups who had participated in these communal, larger-scale projects (Ankeny & Leonelli 2015).

The use of community databases made it possible to dramatically increase the quantity of information on model organisms that could be stored and integrated, as well as the quantity and geographical spread of researchers with access to such information. This quantitative shift brought about a series of qualitative changes in the nature of the community that could do work with a particular model organism and the ways in which members of such communities could communicate with each other. Community databases were established to enable researchers to locate information on a given organism without having to read through all existing literature or be personally acquainted with all research being done. They also fostered researchers' abilities to move across biological subfields at a time when biological research was increasingly fragmented and specialised, for instance, by facilitating searches on state-of-the-art insights coming from different subfields and exposing researchers to diverse instruments, methods, terminologies, and standards. As a result, model organisms became accessible well beyond the original communities that established them and became extremely powerful tools for biological research.

4.5 Material Infrastructures: Stock Collections

In parallel to the growth of data infrastructures for model organisms, the production, use, and dissemination of actual specimens of these organisms has been increasingly standardised and centralised, in part due to large-scale funding out of recognition of the importance of such initiatives. Stock centres have been established in order to collect, maintain, and ensure access to strains of specimens that proved to be particularly responsive within laboratory settings. Since their early days, prominent labs within each model organism community have established and circulated protocols and standardised methods about how best to handle organisms in the laboratory, so as to ensure continuity in experimental procedures and results. The Morgan laboratory started to classify and standardise *D. melanogaster* specimens and distribute them to other labs in the 1920s (Kohler 1994); various strains of JAX mice have been produced and disseminated by the Jackson Laboratory since the 1930s (Rader 2004); *C. elegans* stocks have been available via a formal strain centre since 1978; and *A. thaliana* collections were established in the 1930s in Germany then moved to twinned stock centres in the United Kingdom and the United States, with the Laibach collection still featuring at the core of *A. thaliana* stocks. These collections became increasingly centralised, with many communities agreeing on one or two key sites to place in charge of storing, maintaining, and disseminating stocks on demand.

Access to specimens from these collections initially occurred via paper catalogues, newsletters, and informal contacts. As digital infrastructure became established, stock collections became increasingly integrated with databases, facilitating the posting of accurate and up-to-date information on stocks and promoting the selection and obtaining of the 'right (strain of the) organism for the job' by interested researchers. Although community databases typically have no direct responsibility for how specimens are collected and distributed by stock centres, they still play key roles in supporting the work done at stock centres by offering centralised online access to specimens (Rosenthal & Ashburner 2002). This service requires tight coordination between the ways in which stock centres describe their specimens and the information reported online about them within the databases. Further, database curators must align information about each available strain of mutants with the online data available in relation to those strains. Because these collaborative activities are essential to the coordination of stock centres and databases to permit systematic choice and use of strains by researchers, the curators of community databases clearly influence the ways in which specimens are described, stored, and disseminated to users.

The organisms *C. elegans* and *A. thaliana* are the most successful examples of close collaborations between stock centres and databases: the Caenorhabditis Genetics Centre is directly accessible through WormBase, while the two existing *Arabidopsis* stock centres (the European Arabidopsis Stock Centre and Arabidopsis Biological Resource Centre) were developed and expanded in the late 1990s in collaboration with TAIR. The fruit fly and mouse communities have generally been less efficient in aligning database development with the standardisation of stocks, primarily because stocks of these organisms have not yet been successfully centralised. In the fruit fly community, collections are greatly diversified and some are privately held. The mouse situation is even more diffuse, as stock collections are highly diverse and mainly held by individual laboratories or institutions. Even (one might say especially) in these situations, community databases play a key role in guaranteeing access to stocks. FlyBase lists all existing *Drosophila* collections, which can then be contacted by users for orders, while mouse collections can be obtained through a portal called the International Mouse Strain Resource. The absence of a centralised stock centre with a direct link to Mouse Genome Informatics is the object of heated debates within the mouse community (e.g., Sundberg, Ward, & Schofield 2009). Some argue that this lack of common access has delayed, and in some cases impeded, research progress (Anonymous 2009).

4.6 Conclusion: Characteristics of Model Organism Research

As detailed in this section, distinct 'model organism communities' have formed around particular organisms in order to benefit from the expertise, methods, instrumentation, and data accumulated by participating biologists. Although these communities have many diverse characteristics – differing in size, internal structures, and degree of formality, among other factors – we contend that they share a set of norms and attributes associated with encouraging collaboration and resource sharing, and these norms are closely connected to the epistemic value associated with model organisms (see Table 2 for a summary).

In addition, model organism research was grounded in the broader landscape present in the 1980s–2000s relating to the HGP within which these projects

Table 2. Characteristics of model organism communities

Characteristics of the Community	*Conceptual commitments*	Evolutionary conservation Holistic, inter-level approach to organisms Focus on organisms in isolation from environment
	Available technologies	Well-developed community databases Fit with available instruments and tools (e.g., sequencing techniques)
	Shared skills and practices	Commitment to free exchange of materials, data, and knowledge Ability to move across biological subfields (and related instruments, terminologies, and standards) Public relations skills in attracting funding and attention from outside the scientific community
	Institutional organisation	Charismatic leaders with strong organisational and scientific skills Efficient and accessible stock centres Common communication venues and institutions (e.g., steering committees, journals, community databases, and organism-focused conferences)
	Dependable funding sources	Long-term support from governmental funding Strategies to secure that funding

were funded, but also was shaped by a range of institutional, social, political, and economic factors, as summarised in Table 3.

We return to the components summarised in these tables in Section 5 where we explore the notion of a model organism repertoire.

Table 3. Characteristics of broader scientific, institutional, and economic landscape of model organism research

Characteristics of the Broader Landscape	*Fit with political and social goals*	Vision of basic molecular research as grounding innovation in medicine and agriculture
	Intellectual property regime	Free or otherwise well-regulated exchange of materials, techniques, and data
	Institutional buy-in	Existence of institutions willing to invest in model organism research (through hires, capital investment, infrastructures, etc.)

5 The Model Organism Repertoire

5.1 Introduction

In this section, we frame the stable alignment of elements underpinning model organism work as an important type of *repertoire* within biology. The robustness of this repertoire is fundamental to how model organisms are made to represent beyond themselves, and to the plausibility of using model organisms as models (as we shall see in Section 6). As we illustrate, its significance is underscored by recent attempts to establish new model organisms, as well its role in facilitating translational research 'applying' model organisms to clinical and agricultural research.

5.2 Establishing a Repertoire

The norms of model organism communities described in Section 4 were developed alongside other changes in the contemporary biosciences, including the increasing professionalisation, globalisation, and computerisation of biological and biomedical research which has been characteristic of the past three decades. As a unique way of doing science, model organism research thus involves a set of characteristic practices that emerged in relation to a set of distinctive epistemic goals, which in turn have been finely tuned to the study of the phenomena that these models are taken to represent. These practices together have culminated in what we call a model organism repertoire. A *repertoire* is a general framework for analysing the emergence, development, and evolution of particular ways of doing science. In a repertoire, the successful alignment of conceptual, material, logistical, and institutional components (including specific skills and behaviours by participants in scientific efforts) results in a blueprint for how to effectively conduct, finance, and support research in the longer term (for more detail including how this account compares to others in the philosophy of science that explore collaboration and related issues, see Ankeny & Leonelli 2016).

Model organism research is an excellent example of a repertoire, inasmuch as it depends on specific material, social, and epistemic conditions under which individuals joined together to perform projects and achieve common goals, in a way that was relatively robust over time despite changes in the broader landscape and other features. In the case of model organisms, the adoption and increasing entrenchment of specific theoretical commitments, such as the working assumption of evolutionary conservation and commitments to emphasis on integrative, cross-level accounts, were key components of the story. Looking specifically at cases of modelling in biology, Griesemer refers to the core beliefs of a research community as its 'theoretical perspective', that is, the set of concepts, interests, and values that are (largely unreflectively) used by biologists in their research and demarcate their epistemic culture. He characterises the goal of a theoretical perspective as 'coordinating models and phenomena' (2000, S348); a perspective is thus responsible for determining which aspects of a theory (and hence of the models collectively constituting that theory) are relevant to phenomena and how. A theoretical perspective thus does not apply directly to a specific set of phenomena. Rather, it contributes the analytic and practical tools needed by a scientific community to pursue and obtain knowledge about a specific phenomenon.

Importantly, a theoretical perspective thus conceptualised is grounded in the adoption and development of specific instruments, techniques, and ways of choosing and handling materials. In the case of model organisms, standardisation procedures for both the organisms themselves, the ways in which data and samples were handled (e.g., infrastructures), and the availability of laboratory techniques and reliance on tools such as high-throughput sequencing machines turned out to be crucial to the development of the repertoire. In addition, broader social, institutional, and financial conditions of research shape what is considered as interesting and valid: these conditions are instrumental in facilitating the adoption of specific models as reference points for scientific work. The popularity of model organisms in contemporary biology has not occurred because they constitute the 'best' materials or models with which to work in any objective sense, biological or otherwise; animal models are never a 'given' (Lewis et al. 2013). Nor did they become so prominent because other species are too experimentally difficult or unwieldy, even though many model organism species were initially adopted because of their tractability, ease of storage, and low costs of production and maintenance. Instead, these species have risen to prominence thanks to their proponents' efforts to portray them as 'obligatory passage points' (Callon 1984) for multidisciplinary collaboration across biological subfields.

Thus, the model organism repertoire is a specific type of system of practice that aligns itself very closely with broader regimes, existing technological and institutional platforms, and existing experimental systems as well as a precise theoretical perspective, all of which combine to allow model organisms to 'represent' for those who utilise them. Thousands of researchers from a variety of locations across the globe came to be involved in enacting and developing a broadly shared repertoire that included the very conceptualisation of specific organisms as models of reference for a large work programme with related theoretical commitments about which research questions to pursue. The repertoire also encompassed strategies to acquire blue-skies funding support particularly from the US and UK governments, which in turn enabled research to develop within relatively well-resourced conditions. Specific norms and behaviours, particularly an ethos of sharing data and techniques prior to publication, were attractive to like-minded researchers and contributed to the continuity of the research efforts and their abilities to accrete over time. Finally, the standardisation and centralisation of the production, use, and dissemination of specimens in stock centres, and the establishment of databases to gather both published and unpublished data in a standardised manner also were critical components. Table 3 provides a synoptic overview of these components as discussed in this and preceding sections.

These components may appear to be disparate, but in fact are closely related and tightly interconnected: they arguably could not function effectively without each other. For instance, norms around sharing would not have been sustainable in the absence of large-scale governmental support enabling individual researchers and laboratory groups to disseminate results and materials efficiently in terms of time and resources. Work relating to *Drosophila*, for instance, suffered a temporary set back with the closure of some of its stock centres in the 1990s due to lack of funds. The situation was remedied through lobbying by the research community, leading the U.S. National Science Foundation (NSF) and the NIH to pull together support for one of the stock centres (unsurprisingly, the one most tightly related to FlyBase, the main *Drosophila* database), and the institution of charges for users to help recover costs (Bangham 2019). The two *Arabidopsis* stock centres, the Nottingham Arabidopsis Stock Centre and the Arabidopsis Biological Resource Center, are similarly dependent on a combination of revenue and governmental funding regularly promoted by lobbying and support from the research community. Without a combination of canny management by the stock centres, databases,

Table 4. Components of the model organism repertoire

Characteristics of the Organism	*Natural or intrinsic*	Tractability in the lab Length of life cycle Fertility rates and ease of breeding Size of organism Ease of storage Size of genome Physical accessibility of features of interest
	Induced/uncovered through experimental interaction and transfer to lab	Mutability of specimens Response to lab environment (food, light, temperature, cages, routine) Availability of standardised strains
	Attributed to or projected onto the organism by researchers	Representational scope (how extensively the results of research with the organism can be projected onto a wider group of organisms) Representational target (number of phenomena that can be explored via the organism) Power as genetic tools Ability to serve as the basis for comparisons to other organisms Multi-disciplinary usefulness (capacity to fit different research domains, e.g., genomics, development, and physiology) leading to cross-level integration
Characteristics of the Community	*Conceptual commitments*	Evolutionary conservation Holistic, inter-level approach to organisms Focus on organisms in isolation from environment
	Available technologies	Well-developed community databases

Table 4. (cont.)

		Fit with available instruments and tools (e.g., sequencing techniques)
	Shared skills and practices	Commitment to free exchange of materials, data, and knowledge Ability to move across biological subfields (and related instruments, terminologies, and standards) Public relations skills in attracting funding and attention from outside the scientific community
	Institutional organisation	Charismatic leaders with strong organisational and scientific skills Efficient and accessible stock centres Common communication venues and institutions (e.g., steering committees, journals, community databases, organism-focused conferences)
	Dependable funding sources	Long-term support from governmental funding Strategies to secure that funding
Characteristics of the Broader Landscape	*Fit with political and social goals*	Vision of basic molecular research as grounding innovation in medicine and agriculture
	Intellectual property regime	Free or otherwise well-regulated exchange of materials, techniques, and data
	Institutional buy-in	Existence of institutions willing to invest in model organism research (through hires, capital investment, infrastructures, etc.)

and their users; public relations efforts by governmental funders across the globe; and regularly updated arguments about the role played by these resources in research development including new data-driven methods, these essential components of the repertoire would have disappeared along with much of the attraction of working with model organisms.

Significantly, the hard-won abilities of researchers to effectively align these components gave rise to a wealth of theoretical and experimental results. It also led to ways of labelling and organising those results for future use for a variety of other research purposes including by those outside the community. This combination of influences shaped biologists' understandings of these organisms themselves. Hence, the community for whom a model organism does productive work (and is therefore accepted as representing a given set of phenomena) defines what counts: the model organism and the community become co-constitutive. The repertoire has been essential to the establishment of model organisms as reference points within biology.

5.3 Emerging Model Organisms

The significance of the model organism repertoire as a way of doing research is underscored by the numerous efforts to establish new model organisms in recent years. A key attraction is the opportunity to implement the repertoire in relation to new biological domains and locations, in the wake of the recognition and prestige associated with work on the original model organisms. In the words of a leading developmental biologist, 'there can be few career outcomes more satisfactory for a bioscience professor than the successful introduction of a "new" model organism' (Slack 2009, 1674–75). Another motivation is to make certain that more traditional model organisms do not displace biological research on other organisms, particularly in terms of funding provided. Researchers who do not work on traditional model organisms often lament being forced to rationalise what they do because work on non-model organisms is viewed as less desirable (e.g., Bolker 1995). For instance, in grant applications that may require defence of the use of something less well established (or explicit plans for sharing and maintaining new organisms, in line with existing model organism norms). Our account of the repertoire associated with model organisms helps to explicate not only what is unique about model organisms, but also why not all biological research can or should focus on them: not all organisms can (or should) be 'model organisms'. The model organism repertoire is one among many repertoires that co-exist within contemporary biological sciences but should not be viewed as a 'one size fits all' answer to what can lead to productive research.

One of the most notable efforts to widen which organisms are utilised has occurred via the introduction of the category of 'emerging model organisms' publicised by the well-recognised Cold Spring Harbour Laboratory book series (Cold Spring Harbor Protocols 2019; for a similar approach focused on 'non-model model organisms', see Russell et al. 2017). A series of protocols details methods, available information and techniques, and the potential utility of certain species of organisms, and is geared at those likely to be unfamiliar with the organism in question. Advocates cite advances in genomics and particularly decreases in the time and costs associated with sequencing, which permit exploration of species beyond those traditionally used. Among those discussed are organisms with long histories of use, such as axolotl (*Ambystoma mexicanum)* for regeneration, development, and evolution; planaria for regeneration, stem cell biology, ageing (see also Valenzano et al. 2017), and behaviour; and maize as an alternative model plant. Others are touted as good comparators to existing model organisms, such as the red flour beetle (*Tribolium castaneum)* for comparison of developmental mechanisms with *D. melanogaster* to address questions concerning the evolution of morphology and other characteristics. Yet another motivation is to identify organisms that might be in some sense 'better' than the existing model organisms, such as a cricket (*Gryllus bimaculatus*), which has a mode of development described as 'more typical' than that of the fruit fly *D. melanogaster*, and a nematode (*Pristionchus pacificus*), which has similar experimental advantages to the nematode *C. elegans* but very different genetics and diverse ecological features.

What is important for our purposes is that these organisms are unlikely to fulfil all of the attributes associated with the model organism repertoire. In many instances, organisms are being selected to explore particular phenomena and not as good models for a diverse range of fundamental phenomena. Hence, researchers are not making the same types of attributions to them as were made to the more traditional model organisms, nor do they share many of the underlying conceptual commitments that grounded model organisms. These changes are partly due to shifts in research interests (and the success of existing model organisms in becoming points of reference for such work) and partly to the lack of infrastructures, established norms, and wide-ranging institutional support for these newcomers. In addition, the broader research landscape differs significantly: the large infusion of resources for blue-skies funding that occurred in concert with the HGP and in alignment with broader socio-technical and political–economic regimes, institutional structures, and moral economies in the 1990s and early 2000s was a highly contingent phenomenon, unlikely to be repeated on this scale. As a result, these research organisms do not function as models in the same sense as traditional model organisms: they do not have, or aspire towards, a particularly wide representational scope or target.

However, it is striking that some of the proposed emerging model organisms *do* incorporate particular aspects of the model organism repertoire in their arguments about why they will be particularly useful for this type of research. For instance, available technologies including genomic methods (in the case of moss *Physcomitrella patens*, and the paramecium *Paramecium tetraurelia)* are cited as advantages. Other organisms including social amoeba (*Dictyostelium discoideum*) are argued to be good choices because of the availability of community resources including a genome database (dictyBase) and stock centre. Thus, some takeaways from the history of model organism development and use are being deployed well beyond their original domains.

Furthermore, it is arguable that some of the original 'model organisms' were not as successful as hoped, or as quickly as was hoped, in part because they were not initially positioned to fulfil all of the components of the repertoire. For instance, chicken was historically an important research organism, particularly for the study of development for which it was extremely tractable given the accessibility and eases of manipulation of embryos, and obviously was very important for commercial reasons within agriculture. However, its use as a model organism faced obstacles, as its genetics are considerably more complex than many of the other model organisms. The initial lack of availability of various molecular genetic techniques, such as transgenics and knockouts, made it difficult to develop. Similarly, although the frog *Xenopus laevis* had a long history as an experimental organism especially for development (Cannatella & de Sá 1993), its genetics were more complex and made less tractable until the development of a diploid organism, completion of genomic sequencing of several species of *Xenopus*, and the advent of gene editing (Blum & Ott 2019).

As can be seen from these examples, what makes model organisms into unique models is the complex intersection of components associated with the model organism repertoire. Not all biologists do or should use model organisms: there are research questions that could benefit from molecular biological approaches that nonetheless may be more appropriate to address in species that are not among the traditional or classic models. As emphasised throughout this Element, model organisms are aligned with a distinct way of knowing, which is not suited to all purposes nor necessarily better than others in any general sense.

5.4 The Translational Role of Model Organisms

One epistemic goal of model organism research that deserves further attention is its role in acquiring knowledge about humans (Schaffner 1986) and developing medical applications such as pharmaceuticals. As noted earlier, model organisms are understood to have a broad representational scope and serve as the basis for articulating processes that it is thought will be found to be common across all (or most) other types of organisms and also have broad representational targets. A particularly important use of model organisms occurs within biomedicine, namely, focus on humans and especially those with some type of disease condition. In those cases, humans constitute the representational scope of the model, and one or more aspects of the disease of interest become its representational target (cf. Piotrowska 2013). Thus, the case of use of model organisms within biomedicine can be seen as a more specific example of the use of model organisms more generally (Huber & Keuck 2013).

Some model organisms have come to be strongly associated with their abilities to allow translation of findings back to humans. These particularly include those on which developmentally relevant genes can be modified (e.g., deleted via knockouts or added via knockins) in a precise and controlled manner, and that are therefore valued as useful predictive models to study human disease mechanisms. The mouse *M. musculus* has been said to be the most commonly used model organism in research on human disease (Rosenthal & Brown 2007), in part because of the large degree of genetic similarity but also due to a range of factors associated with its tractability and long-established track record of associated resources, tools, and technologies. For example, the so-called p53 knockout mouse relies on a tumour suppressor with important functions in relation to DNA transcription, cell growth, and cell proliferation, and hence associated with cancer prevention. Large numbers of human tumours contain mutations or deletions relating to the p53 gene, and hence mice with this gene knocked out have been extensively used to study a range of human cancers, including their proliferation and spread (Davies 2013; Nelson 2018). Zebrafish (*D. rerio*) also have a high degree of genetic similarity with humans, with many conserved genes, pathways, and features, and they are cheaper and easier to raise in large quantities. Although they are not useful to study human diseases in those tissues that they lack (e.g., lungs), knockout *D. rerio* have been used to study the severity and progression of human diseases such as Duchenne muscular dystrophy.

In recent years, concerns have emerged about the lack of replicability of biomedical research using mice. Our account of model organisms helps to provide guidance about these issues. First, the adoption of model organisms can have potential negative side effects, including canalisation with regard to what organisms are utilised for research. There is no doubt that *M. musculus* has in many senses become the default experimental organism for biomedicine. Yet, there may be evolutionary and other reasons to think that use of mice as model organisms is in many cases not well-supported in evolutionary or other terms (e.g., Bolker 1995, 2009; Perlman 2016), which may in turn raise problems with regard to reproducibility.

In addition, as with any other form of experimental research, a series of background commitments must be in place in order for such models to be valid. To be used to model a human disease, the research organism must (in some sense) evidence the same attributes as the condition in humans in terms of the underlying genetic, physiological, or other processes, and in terms of the disease phenotype produced. Hence, certain types of disease conditions simply may not be able to be accurately modelled in mice or other non-human model organisms (e.g., Ransohoff 2018). Moreover, the selection of the allele and strain on which to focus must be made carefully, particularly given limitations of certain strains including genetic drift over time; the environment in which experiments are done must be articulated, and appropriate experimental controls established. What constitutes an appropriate control can be problematic because of the different understandings of the idea of 'control' within the biological and biomedical research contexts (Güttinger 2019), which are themselves related to use of different research repertoires. While biomedical replicability presupposes randomisation as the gold standard, for example, biological replicability focuses more on the accurate reporting of experimental conditions and the investigation of variation as an important phenomenon in and of itself which may help to explain divergent results arising in attempts at replication (Leonelli 2018).

The dominance of mice models in biomedicine points towards an intriguing tension: on the one hand, the mouse *M. musculus* could be viewed as the best example of a model organism, given its prominence and sheer quantity of usage in contemporary scientific practice. However, it arguably does not share many of the attributes associated with the model organism repertoire outlined previously, due to the complexities inherent in the diverse research programmes using mice. The use of research organisms in translational efforts has become a strong motivator particularly within model organism-based research, especially given the considerable amounts of global funding dedicated to biomedical research in comparison to exploratory biological research. Mouse research has been largely commercialised for nearly a century (Rader 2004), with various strains of mice available for purchase rather than via sharing of specimens and similar. Sharing of data has been considerably restricted due to the sizable investments made by private companies as well as public institutions, all of which are in competition with each other, especially in terms of potential pharmaceutical products, thus creating a situation in marked contradiction to the 'share and survive' ethos. As a result, there is no unified community (and only limited discrete subcommunities) of mouse researchers and few central-ised resources. Perhaps most importantly, mice are conceptualised more as tools, with much less emphasis on multi-level modelling or integration of knowledge across levels, disciplinary approaches, and locations in order to understand the organism taken as a whole.

5.5 Conclusion: Using the Repertoire Framework

This section demonstrates the significance of understanding model organism research as a repertoire, as well as the complexity of applying this framework across different communities that invoke the 'model organism' label. Clear articulation of the elements of the model organism repertoire makes it possible to explore similarities and differences between model organism communities as well as the evolution of their practices over time. It also enables the study of variations due to different degrees of exposure to translational endeavours, commercialisation, and regimes of intellectual property, which is important as restrictions on the sharing of results and international collaboration directly impact the effectiveness with which the model organism repertoire can be used to enable an integrative understanding of organisms. It is not enough in this case to assert that many researchers working on mice tend to have different goals from researchers working with the nematode *C. elegans*: even when goals may be the same (e.g., studies of cell degeneration with a view to oncological applications), the differences between uses of these two models involve values and norms of scientific communication, availability and management of infra-structures, and relations between researchers working on different projects. All of these components are intertwined and result in divergences in the researchers' resources and practices. In Section 6, we explore how this approach can inform a deeper understanding of the plausibility of model organisms as biological models.

6 When Are Model Organisms 'Good' Representations?

6.1 Introduction

Starting with a discussion of the importance of understanding modelling as an activity, this section examines what makes model organisms plausible as bio-logical models. In other words, we look at the grounds upon and conditions under which groups of researchers in a community commit (implicitly or explicitly) to the view that model organisms can represent other organisms. Reliance on model organisms as plausible models that are 'good' or 'useful' depends on their tight association with a powerful and effective repertoire. We illustrate these issues through discussion of what was involved historically for the mouse *M. musculus* and the thale cress *A. thaliana* to become accepted as plausible models in alcoholism studies and plant biology respectively. We then discuss the limits and constraints associated with this way of modelling, and address questions arising in cases where model organisms are widely recognised as inadequate and therefore implausible models.

6.2 From Models to Modelling Activities

Models come in an endless variety of forms. A combination of these is always required when they are used in scientific practices, where they interrelate and work together in a variety of ways (for discussion and examples, see the volumes edited by Morrison & Morgan 1999; de Chadarevian & Hopwood 2004; and Laubichler & Müller 2007; as well as Green 2013). Given this dramatic diversity, and the resulting 'promiscuity' (Griesemer 2004, 436) of the notion of model itself, much attention has been paid, particularly by philosophers, to the actual features of models employed in scientific practice. These discussions have allowed clarification off the epistemological status of various types of models (ranging from scale or toy models, to physical or theoretical, to idealised and fictional, and so on: for a summary see Frigg & Hartmann 2018) as both products of scientific practices and as tools used to develop or interpret theories.

Much less attention has been devoted to the variety of activities that need to be performed in order to yield adequate or 'good' models and to productively employ them in processes relating to explaining the world and intervening in it. Scientists not only refer to models in their explanations, but also use, manipulate, and constantly modify them in order to achieve and justify those very explanations, and develop related strategies for experimental intervention. The adequacy of any instance of such use is determined both by the features of the phenomena under scrutiny, and the material, social, and institutional settings associated with the scientific practices at issue and the commitments of the researchers involved. In this context, 'representation' is not taken merely as some sort of complete or static mirroring of a phenomenon, system, or theory by the model. Rather, representation is a type of rendering that is necessarily active and partial, insofar as it is grounded in and instantiated by research practices, which in this case include the development of a repertoire with specific conceptual, material, and social components. Models owe their representational power to this complex set of historical and epistemic circumstances; in this sense they mediate between theory and world (and thus our account is indebted to Morrison & Morgan 1999 and Morgan 2012).

Examining modelling *activities*, rather than solely focusing on their products such as theories, is a useful approach. Such an approach is particularly valuable when investigating how model organisms help to create knowledge that can be projected beyond the immediate domain in which it has been produced and thus serve as 'good' models. Models can fulfil many diverse functions, including as representations, but only by being used, manipulated, or put to work, and is supported by the growing philosophical literature on practice-focused accounts of modelling (e.g., Knuuttila 2011; Chang 2012; and Gelfert 2016 to name just a few) in which our account is grounded. The critical question here is what makes such projections more (or less) plausible, particularly because model organisms are sometimes used to represent phenomena that are arguably not directly observable using the organisms themselves or for targets (such as humans) that are very dissimilar to the original organism. In the terms presented in Section 3, the issue is how to account for the imputation of properties of the model to the target. Such imputation involves understanding communities of researchers' choices and justifications of a key to connect properties found in the model to properties imputed to the target, which in turn determines the representational power of the model, as well as the commitments involved in deploying it in experimental practice.

6.3 Focusing on the Plausibility of Models

Adoption of a 'science in practice' approach requires focus on modelling as an activity that occurs in a complex research environment that includes conceptual commitments, specific shared practices and experiences, institutional framework and aims, and other broader aspects of scientific practice. What makes an organism-based model plausible as a representation depends on the degree to which communities of researchers deem the use of the organism as a model for a given phenomenon to be epistemically fruitful and justifiable within the broader research environment. This notion of plausibility is necessarily dynamic, encompassing a spectrum that can vary from low to high plausibility; it evolves and iterates as additional evidence is gathered, conceptual commitments and practices change, and so on, and is grounded in communities of researchers' overarching perceptions and evaluations of their own research practices and goals.

We thus define *plausibility* as the degree to which communities of researchers deem the use of an organism as a model for a given phenomenon or group of organisms to (1) be acceptable to others, in the sense of being taken seriously as a tool for scientific work by at least some of the researchers' peers (usually a sizeable group spanning multiple locales and institutions); and (2) fit within an epistemic space (Rheinberger 1997, 2010) created by the availability of background knowledge, questions, concepts, technologies, methods, data, and/or materials that researchers are already investigating or using in their work, and which therefore makes it possible for researchers to rationalise and justify commitments to a specific organism. Consider, for example, the case of a researcher interested in a particular type of coral reef because she loves working in tropical regions. The preferences for spending time in a pleasant location and the affective link between the researcher and certain organisms are important grounds for choosing them and likely for continuing to use them (Ankeny & Leonelli 2020; Dietrich et al. 2020), yet do not contribute towards making the organism in question a *plausible* model. To do so, the researcher must consider how the organism can contribute to ongoing debates in her field, to the collection of data relevant to a specific set of questions or phenomena, or to the development of techniques or technologies that may foster scientific innovation, in conjunction with the features of any repertoire that might exist in her research community. These arguments frequently occur in different types of contexts, ranging from lab group discussions, grey literatures such as informal community communication mechanisms, and conference presentations, administrative reporting, funding applications, and publications. But as observed in the case of emerging model organisms (Section 5), only once that type of argument is made, and at least some peers are persuaded that the choice of the organism is acceptable as a model, can the researcher with some degree of confidence use the organism explicitly as a model in her research. She may of course do research without making these types of modelling-related arguments, especially in early stages where information is being gathered and the potential for the organism to serve as a model is being investigated, all of which contribute to future arguments about plausibility.

There are several qualifications necessary in relation to this definition of plausibility. First, we are well aware that the size of the group of peers that find the model acceptable will vary enormously depending on field, type of organism, time period, and a range of other factors, and hence we are not wedded to any specific community size in relationship to the acceptability of a model. For our purposes, we only wish to signal that the plausibility of a model necessarily involves some degree of social consensus beyond the preferences of a solitary investigator (who of course is largely a fictional figure within today's extensively collaborative and team-based research).

Second, the idea of acceptability does not necessarily mean that the use of an organism as a model is empirically well-warranted or successful using any definitions of these terms. A given model may well be viewed as plausible before much or any substantive evidence is produced about it or knowledge claims established through its use, but still fit within the available epistemic space that constitutes the second half of our definition discussed earlier. Again, there is undoubtedly a continuum present here in terms of available empirical information, and the processes that occur to strengthen arguments about plausibility. Consider our earlier historical discussions, where for example arguments about serving as a model were thought to be quite compelling to a particular community of researchers in the earliest days of research using the nematode *C. elegans* even though actual evidence about some key features (e.g., degree of genetic conservation) was in fact quite limited. Relatedly, the extent to which these arguments are explicit in its strictest sense (e.g., in published peer-reviewed literature) may well vary, again according to the fields, research cultures, historical moment, and so on.

Finally, we recognise that operationalising this definition by developing a defined metric of plausibility (such as for instance how many peer-reviewed journal articles make arguments or cite a certain organism as a good model) will not be possible. Any such metric would invariably be linked to a very specific set of assumptions around the characteristics of 'successful' research, which can vary widely across research cultures in both space (field, geographical, or otherwise) and time (historical or stage of research programmes). Thus, while there may be scientific and sociological concerns around whether our definition can be operationalised, what we wish to emphasise are the significant advantages that a focus on plausibility provides which in turn allow philosophical understanding of the use of organisms as models in biology.

What are the advantages of our focus on plausibility? For a start, the dynamic nature of this notion requires us to explore the idea that a model can gain more (or less) plausibility over time depending on a range of issues including, but not limited to, its representational power. Researchers typically engage in processes of 'plausibility building' (see Hoffer 2003 on a similar idea in medicine), which involves gathering a range of types of information and considerations to make the use of a particular model more (or in some cases less) plausible. So in the case of our hypothetical coral reef researcher, although she may start working on her organism of choice for a range of affective reasons, her research may permit her to gather evidence, try out new techniques or methods, investigate available data from other domains in the context of her organism, and so on, which can contribute to plausibility building and result in a compelling argument about the use of the chosen organism as a model (relatedly see Weber's idea (2005) of 'preparatory experimentation'). In addition, this notion of plausibility grounds determinations about what makes a particular model choice a good one, including considerations of researchers' overarching perceptions and evaluations of their own research practices and goals.

The notion of plausibility has been used in other contexts, notably in the so-called Bradford Hill criteria associated with epidemiology and medicine, where it is used to evaluate the reliability of causal claims. It is useful to contrast our notion to this one: the focus in the Bradford Hill criteria is primarily on biological plausibility in that it holds that proposed causal associations must be consistent with existing biological and medical knowledge including theoretical commitments and working assumptions; hence it is analogous to the focus on mechanisms in the philosophy of biology literature (as discussed in more detail in Section 6.4), and particularly genetic and other forms of evolutionary conservation as the main (or sole) grounds by which researchers make arguments about organisms as models. By contrast, we wish to consider the broader conditions under which such judgements are made, and thus are interested in which factors underpin researchers' determinations with regard to plausibility, ranging from specific investigative experiences to broader institutional frameworks and aims.

Our notion of plausibility is also distinct from several other concepts that are nonetheless relevant but not co-extensive with it, and which have been dominant in philosophical scholarship on models. The idea of *credibility*, for instance, is heavily evidence based and focuses on empirical warrants for imputations of plausibility, rather than the broader set of considerations that may ground judgements of plausibility beyond the presence of empirical data. Another related but distinct concept, *realism*, refers to the capacity of a model to capture or reflect aspects of the world in a truthful or accurate manner; although such considerations might be relevant in some cases, this concept does not help us to understand cases where highly idealised or modified organisms are taken as plausible, though not realistic, models. Many philosophers also discuss the representativeness of organisms as models, in ways which are either too vague or require a close mapping of features using similarity relations or isomorphisms (see our critique of this approach to representation in Sections 3.4 and 3.5). We think that plausibility provides a much better entry point for understanding how representational claims are made within actual scientific practices, in all of their rich complexity, and we return to this issue in Section 7.

6.4 Plausibility in Action: Establishing Good Models

To date, philosophers' assessments of what makes a 'good' model for research (or in our terms what makes it 'plausible') have largely relied on articulation of the underlying mechanisms relating to the phenomenon of interest as providing the basis for them to serve as representations (e.g., Cartwright 1989; Schaffner 2001; Weber 2005; Craver & Darden 2013). Models that share mechanistic features with their targets are more likely to generate the phenomena of interest via the same causal pathways and respond in similar ways when these pathways are disturbed. Hence, shared underlying mechanisms is argued to be the basis for establishing or determining what we term the 'plausibility' of a model; to put it in Frigg and Nguyen's terms, it is taken to provide the key through which properties of the model are identified as relevant and imputable to the target.

In the case of model organisms, such arguments are undoubtedly an important part of what underlies assessment of plausibility, particularly given typical implicit assumptions about genetic and other forms of evolutionary conservation. Many have claimed that inferences from model organisms are empirical extrapolations based on evidence about shared genetic ancestry (e.g., Steel 2008; Bolker 2009; Weisberg 2013; Levy & Currie 2015). Such extrapolations involve treating the organism as a representative specimen of a broader class (e.g., as Bolker 2009 terms it, a 'surrogate'). Such accounts are insufficient to understand what is occurring in research practices that depend on model organisms for at least two reasons. First, these views rely on narrow definitions of what counts as 'extrapolation' (or the related concepts of 'interpolation'), arising largely from statistical and quantitative ideas about using known relationships that have been measured and observed to understand those that have not yet been (see also Baetu 2016). Second, they do not facilitate the analysis of why and when model organisms are viewed as (more or less) useful and appropriate to use in actual scientific practice. Traditional accounts are overly focused on the underlying entities and their attributes without attention to what is done using these entities during research-related processes, which in turn tends to reinforce static ideas of what counts as a good model. Our account relies on attention to modelling processes in action where judgements about models, and the commitments that underpin their uses as representation, are highly iterative and evolve over time. In other words, these processes are brought about through changes in understandings relating to components of the key used to connect properties of the model to those imputed in the target.

By way of illustration, we now briefly consider condensed histories of two models that ended up being widely adopted despite being initially judged as 'implausible': *A. thaliana* for plant science (summarised from Leonelli 2007b) and *M. musculus* for alcoholism research (drawing on Ankeny et al. 2014).

6.4.1 Making a Model Plant: Arabidopsis thaliana

The first era associated with interest in *A. thaliana* as an experimental organism occurred in the 1940s, mainly through the work of Laibach based in Frankfurt' and Rédei at Columbia University. Laibach was drawn to the species due to its extraordinary natural variation in phenotype, and began the first systematic collection and classification of *A. thaliana* wildtype mutants in the late 1930s.

He believed that *A. thaliana* could become a suitable organism to study the mechanisms responsible for its surprising diversity due to its tractability, accessibility, short generation time, and relatively simple genetics.

Upon his retirement in 1965, Laibach's wildtype collection was replicated and shared so that researchers in Europe and the United States could access the full complement of *A. thaliana* lines. At around this time, Gerald Röbbelen began publication of the Arabidopsis Information Service, an annual newsletter that gave updates on experimental work and improved lines of communication. In his first editorial, Röbbelen billed *A. thaliana* as the 'botanical *Drosophila*'. However, early attempts to bring the plant into the laboratory had not been successful at this point for two main reasons. First, most post-war plant research was focused on breeding techniques, particularly on agriculturally significant organisms such as tobacco, which meant no funding was being provided for plants viewed as commercial dead ends such as *A. thaliana*. Second, *A. thaliana* appeared to be resistant to chemically induced mutation, a method that was proving highly effective in other organisms: the few artificially induced mutations obtained between 1950 and the early 1980s were costly, requiring months of experimental labour with no foreseeable hope of speeding up the process. The resilience of *A. thaliana* to genetic modification caused many European and American biologists to turn away from it to more tractable model organisms such as baker yeast (*Saccharomyces cerevisiae*), tomato (*Solanum lycopersicum*), and maize (*Zea mays*).

In 1986, an unexpected solution to this obstacle emerged in the form of a simple technique for generating mutants: spraying the *A. thaliana* wildtype with a bacterium (*Agrobacterium tumefaciens*) doctored with a plasmid permitting incorporation of a gene of interest. Thus it became suddenly possible to easily obtain a great variety of *A. thaliana* mutants in which phenotypic growth had been disrupted. This development won a great number of biologists over to the study of *A. thaliana*. The organism also was explicitly marketed to animal geneticists as offering the opportunity to pursue new lines of enquiry in a less competitive environment, one free from the dogma of an older generation and where research could be organised afresh in a highly collaborative fashion. *A. thaliana* had a rich and well-systematised collection of ecotypes and a simple chromosomal structure; even better, some genetic data were already available and yet it had not become the subject of a large research project. The relevant question for molecular biologists in the 1980s became: 'why *not* a plant?' Several meetings were organised to answer this question, giving shape to a community of likeminded researchers. It thus became acceptable to study a plant that was highly suited to laboratory life, but of no immediate agronomic interest.

In the mid-1980s, the NSF decided to provide abundant funding to the US proponents of *A. thaliana*. Lobbying by the molecular biologist James Watson (who later was the Head of the US part of the HGP at the NIH) certainly contributed to NSF's benevolence, as did its desire to enhance its profile among US funding agencies. Outside the United States, researchers argued for the need for similar investment to avoid being left behind, leading other governments to quickly follow the NSF's lead. Within five years, Britain, Germany, the European Union, and Japan were ploughing considerable amounts of money into *A. thaliana* research. In 1990, the Arabidopsis Genome Initiative was born, a multinational research effort that successfully managed to yield a complete map of the *A. thaliana* genome by the year 2000. As the first plant to undergo complete genomic sequencing, its preeminent status within plant biology was confirmed.

6.4.2 Using Mice to Model Alcoholism

Mice, particularly *M. musculus,* have long been well-accepted models particularly for behavioural investigations, given that they are highly standardised, readily available via stock centres, and easily manipulated under experimental conditions. But use of non-human animals in general has been contested when it comes to alcoholism research, where free will and human volition, as well as a complex set of behaviours and social relations, seem integral to the study of the disease. The successful adoption of mice as a plausible model for twentieth-century alcoholism research thus involved considering not only the features of the organisms themselves but also the environment and experimental settings within which they were studied.

Early alcohol researchers tended to work with a diversity of types of organisms and to use a wide variety of experimental set ups. However, one of the major problems increasingly recognised by researchers in later part of the twentieth century was how to use animal models to understand the uniquely human phenomenon of alcohol addiction: even when interested in using some alcohol, non-human animals do not tend to consume large amounts of it, and mice and rats are especially reluctant to drink alcohol when given a choice between alcohol and water. Thus, alcohol addiction in experimental animals is generally viewed as induced, inasmuch as the behaviours and preferences of animals have to be transformed in order for them to serve as experimental models for human alcoholism.

Hence, the community of alcoholism researchers developed detailed criteria for the methods and experimental set ups required to make them plausible as models: animals have to self-administer alcohol by the oral route and consume it in quantities that would result in pharmacologically significant blood alcohol levels; alcohol should be consumed for its pharmacological properties and not for its taste or caloric properties; animals should be willing to work for alcohol; and tolerance and dependence must emerge as a result, measured by reduced effects of alcohol consumption and acute withdrawal symptoms (Cicero 1979). These criteria became a touchstone for later debates in the field of alcohol addiction research about the plausibility of certain animal models and helped to ground key features of the experimental set ups on which the community came to agree, namely, the characteristics of the cages in which organisms are kept and the actions they are required or allowed to perform (Ramsden 2015). It is critical to note that researchers relied heavily on the experimental set ups (rather than underlying conceptual commitments, for instance) in part to avoid making firm commitments to what causes alcoholism and the extent to which it depends on environmental factors, a question that is a key part of what their research aims to answer. In addition, the adoption of mice in this context has involved a narrowing of research focus amongst some researchers to genetics and physiology (see Nelson 2018 for more detail), separate to an extent from social and behavioural factors which are studied by others using human experimental subjects.

6.5 Model Organisms as Plausible Models

Plausibility judgements thus shift over time: there is no one 'good' model organism at all points in time for all purposes. The brief histories provided earlier highlight the ensemble of conceptual, technological, and social developments that facilitated the adoption of these two organisms as plausible models where there were initially viewed as rather unlikely. In both cases, repertoires play critical roles in shaping the interpretations that are required whenever a model is taken as representing. In other words, some of the components of the repertoire provide the key through which properties exemplified by models can be associated with properties of their representational targets (Frigg & Nguyen 2018). These components include conceptual commitments (particularly the implicit working assumption of evolutionary conservation and genetic/genomic approaches as the primary focus to the exclusion of experiments involving environmental factors), availability of methods (e.g., *Agribacterium* transformation to produce *A. thaliana* mutants) or consensus about experimental approaches (e.g., Cicero's criteria as a way to control mice behaviour and generate the 'right kind' of responses to the environment), and explicit lobbying with funders and institutions (e.g., *A. thaliana* researchers' appeals to national prestige). So long as these features were not in place, the use of the organism as a model was not thought by researchers who were familiar with it to be justifiable.

We can now enrich the illustration of how model organisms represent presented in Section 3.5 (Figure 1) to include the role of repertoires in relation to the key.

The communities utilising model organisms share in the experimental methods and conceptual commitments underpinning them, which constitute the key for using those organisms as models. For instance, they agree that the organism's environment is not relevant, inasmuch as external environmental conditions have been black-boxed through processes within the laboratory that hold them constant in the form of a highly reified experimental setting; they also have come to assume use of a highly standardised organism so that it can be assumed to be the same from lab to lab and over time. Whether a model organism is a plausible model also hinges on the availability and access to the phenomena of interest, and a range of infrastructures that provide models, diagrams, and various types of information in a usable and interoperable format (Leonelli & Ankeny 2012) that permit connections to be made between the model and its intended target. Thus an organism's ability to represent any particular phenomenon is only partly determined by material features of organism itself (including the degree of evolutionary and particularly genomic conservation): many other factors are involved in the broader scientific practices associated with the organism's use, and contribute to researchers' judgements about how likely it is to be a plausible model.

6.6 Limits of Model Organisms as Representations

This exploration of how model organisms come to be viewed as plausible models and the roles of the key and the repertoire in how they represent allows us to reflect on cases where such models are viewed as implausible. For instance, it has been claimed that reliance on a small number of model organisms does not permit adequate understanding of the relevant phenomenon relating to biodiversity and development (Bolker 1995; Bolker & Raff 1997). Relatedly, evo-devo research is said to use criteria for the selection of experimental organisms that differ greatly from those criteria used to select suitable organisms for molecular studies including on model organisms (Jenner & Wills 2007; Sommer 2009; Minelli & Baedke 2014). Hence, for questions associated with environmental influences on development or wider understandings of natural biodiversity and evolution, critics note that model organisms are likely to be limited in their abilities to serve as models. Relatedly the processes of laboratory-based cultivation associated with model organisms have involved idealisations or known departures from features present in the model's target (Ankeny 2009); for instance, most of the traditional model organisms share the same developmental processes, including immediate separation of the germline from somatic lines (Gilbert 2001). Thus for research focused on diversity in these developmental processes, model organisms are likely to be implausible or 'bad' models. We can usefully view these claims as related to different interpretations of the key (as well as in some cases to distinct targets). Researchers whose primary focus is on certain types of biological phenomena (e.g., environmental influences on development or questions in evo-devo) do not believe that they will be able to impute the properties of the model (model organisms) to the target of their studies in a manner that will allow them to answer their research questions.

Thus, although there is no doubt that these concerns are well-founded, our account of what makes for a 'good' model aims to go beyond the sorts of features which relate only to empirical findings arising from research, particularly those kinds of claims which biologists produce and defend with reference to specific organisms (as stressed for instance by Weber 2005). While there will be cases where empirical or biological information (e.g., claims about conservation) may well be the main grounds for arguments about plausibility (and hence constitute the critical property within the key), these claims are always made within the context of the conditions associated with our definition of plausibility. Thus, empirical claims are always part of a broader epistemic space which will have repercussions for any broader uses of an organism as a model. So, for instance, criticisms of use of model organisms as developmental models rely on empirical claims about typicality, particularly in comparison to natural or wild organisms. However, they are often embedded in a broader epistemic space which does not share many of the attributes of the model organism repertoire, for instance, the privileging of use of molecular techniques, and also frames its concerns in terms of different types of questions than those typically pursued in model organism research.

The model organism repertoire provides an excellent, explicit, and replicable way of grounding claims about model plausibility. Even in cases where the similarity between the organism and its target seems obvious, researchers still need to make explicit arguments to their peers about why they should accept an organism as a model. Hence, it is useful to reflect on the properties of the model that can legitimately be attributed to the target and therefore used to ground arguments about plausibility.

At the same time, the model organism repertoire may inadvertently warrant a kind of 'organism imperialism' (see also Hopwood 2011 on the related idea of 'species politics') by making it much easier to stick to one of the traditional model organisms rather than investigating whether other organisms may be better suited to the questions and research environment at hand (Ankeny 2010). Components of the model organism repertoire may be employed as part of the key to justify uses of model organisms as models in situations where they are in fact less appropriate: in some sense, the use of rodents as models for alcoholism, as discussed earlier, potentially provides an example of introducing a range of what some might view as ad hoc experimental set ups and conceptual commitments in order to maintain a genetically focused research programme. But as noted in Section 5.3 with regard to emerging model organisms, the model organism repertoire will not be valid or useful in many research contexts and hence should not be forced or assumed.

It also has been argued that existing limitations of model organisms can be in some cases converted into strengths, if the problem or phenomenon to be studied is carefully selected (e.g., Jenner & Wills 2007). For instance the nematode *C. elegans* is well-recognised as extremely developmentally stable and resilient even in face of environmental and other types of perturbations, which makes it very dissimilar to many other organisms. However, as a consequence, worms can be used to study the evolution of this type of developmental canalisation which in turn might permit correlated features and their underlying mechanisms to be articulated and their absence to be explored in other species. Thus, seemingly 'biased' models can permit biologists to address important general issues, and hence 'represent' even in these sorts of extreme cases, again so long as arguments about plausibility are well-grounded in clear articulation of the components of the key.

6.7 Conclusion: Good Models for What Purposes?

This section has defended the crucial role of the model organism repertoire in underpinning researchers' arguments about the plausibility of model organisms as models, with particular focus on the use of a key closely associated with model organisms. As we have stressed, it is critical that model organisms are only 'good' models in particular contexts, and hence it is important to examine for whom they are 'good'. As Weisberg (2007) has noted in asking 'who is a modeler', researchers' goals are key to establishing any model and legitimating its use. While agreeing with him on this important point, the roles of institutions, regimes, instruments, skills, and political economy are critical. Therefore, our view is broader and more cognizant of a range of scientific practices than what his focus on individual preferences appears to admit as relevant. The brief cases of *A. thaliana* as a model plant and 'alcoholic' rodents as presented provide clear evidence about the need to explore these types of factors: the justifiability of use of these changed over time, and they would not have been widely adopted without various forms of institutional legitimisation, explication of shared conceptual commitments, and technological developments.

In closing, we note that our notion of plausibility has some overlap with Weber's claims about 'vindication', which he describes as occurring in later stages of organism-based research. He notes that model organisms only 'prove their worth after a while, by enabling fruitful research in many different laboratories, the results of which can again serve as a basis for further research' (2005, 179). This claim highlights the iterative nature of the process in parallel to our account illustrated in Figures 1–2. He argues that the generality of certain biological principles over a large number of species is not inferred by enumerative induction, but through a more sophisticated type of inductive argument which relies on phylogenetic conservation. He also claims that model organisms have epistemic functions over and above providing a basis for inductive inferences or extrapolations to other organisms, including use as important tools (see Germain 2014 on instrumental uses of animal models more generally). Our analysis underscores these considerations and provides more detail about the processes associated with these types of practices by placing them in a broader context. This context involves the intertwining of funding and institutional structures, technologies and techniques, community dynamics, and effective marketing with conceptual and methodological commitments, and biological or material factors.

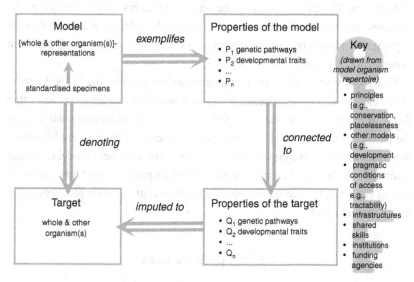

Figure 2 How model organisms represent, including the essential role of repertoires as underpinning the choice and entrenchment of the key

7 Conclusions: What Future for Model Organisms?

7.1 Introduction

The special role of model organisms in contemporary biology is undeniable. What we have added through our account is a detailed argument about the underlying mechanisms and rationales associated with the science that is practised using them, with particular focus on the epistemological implications related to use of these organisms. The ability to sequence the genomes of key species clearly magnified their role as platforms for integrative biological research. By the end of the last century, *C. elegans, E. coli,* and *A. thaliana* had become the first organisms to have their DNA fully sequenced, thus opening an entirely new space for research aimed at deciphering the code, which although now detailed still requires considerable interpretation. The importance assigned to implicit assumptions about evolutionary conservation, the vision of genomic sequencing as an indispensable tool for biological research, and other commitments, clearly affected the choice of model organisms: for instance, species whose small genome was more amenable to detailed molecular analysis (such as *A. thaliana*) were favoured over species with big genomes. For example, even though maize (*Z. mays*), was very popular as a research organism in the 1950s and 1960s because its large chromosomes are visible under the microscope, it has a large genome that was not as tractable for sequencing-based approaches. But closely related to these sorts of technological and material factors were a range of institutional, social, and conceptual elements that contributed to shaping model organism research, as documented throughout this Element. These in turn affected the ways in which research centred on model organisms has been performed, with commitments linked to the adoption of model organisms as models becoming entrenched over time in biologists' approaches to experimental practice.

With the advent of new sequencing technologies, genome size is less likely to affect the choice of which organism is best suited to experimental and other forms of research in specific areas. Other factors, such as natural variability or the potential environmental or commercial impacts of a species, are becoming more prominent. In this concluding section, we discuss where model organism research is likely to be heading, including both its prospects and limitations, as well as exploring broader emerging trends within contemporary biology arising out of such research.

7.2 Comparative Uses

The use of model organisms as reference for comparisons across very diverse species, including some that are very different from the models themselves, is frequently highlighted as a key trend in contemporary biology and one of the main reasons for the continuing importance of model organisms. The ability to use model organisms in this way has been strengthened by the availability of high-throughput data production and increasingly sophisticated computational methods for data storage and analysis. These include the construction and alignment of reference genome networks (e.g., Srinavasan et al. 2007); the development of powerful data warehouse systems, such as Intermine (Smith et al. 2012), created specifically to integrate and analyse complex biological data, and historically grounded in long-standing attempts to link model organism databases; the construction of standardised, crowdsourced annotation systems for reference genomes, such as the Chado system (Mungall et al. 2015), through which researchers can contribute and compare data, metadata, and background knowledge; and the use of programming interfaces to match genotypic and phenotypic profiles that facilitate comparisons between model and non-model organisms, as well as between model organisms and humans for clinical purposes (e.g., Mungall et al. 2015).

It should not be surprising that model organisms continue to play important epistemic roles in this context. Comparative approaches are central to how model organism researchers have conceptualised and carried out their work over the last fifty years (see Section 3.3). Far from being a new component of model organism research due to the availability of new technologies as claimed by some authors (e.g., Strasser 2019, 259), the opportunity to compare model species with more diverse types of organisms has always animated this way of doing research and constituted one of its key goals (Ankeny & Leonelli 2011). Throughout the 1980s, 1990s, and 2000s, one finds a plethora of scientific publications exploring opportunities for comparison not only among specific strains of model organisms, but also with similar species, frequently within the same family. Thus, on the one hand, model organism research aimed to perform comparisons across similar organisms as a first step towards broader comparative exercises and in order to work out which tools, knowledge, and resources were actually needed to fulfil this aim. On the other hand, the attempt to develop tools, standards, and infrastructures to facilitate comparison among wildly different species has also underpinned the history of model organism research from the beginning, including a focus on understanding the conditions under which integration of data coming from different model organisms may be possible and informative. Several joint initiatives occurred in the 1990s–2000s, which brought together representatives of different model organism communities to devise ways of sharing data and comparing insights. These initiatives included, for instance, the development of a Gene Ontology embracing all major model organisms and of a Generic Model Organism Database toolkit through which data could be linked, visualised, and analysed together (Leonelli & Ankeny 2012; Leonelli 2016).

This build-up of expertise and resources facilitating comparative analysis resulted in an explosion of papers using model organisms to understand features of what are frequently characterised as 'non-model' organisms. For example, the thale cress *A. thaliana* is used to study vernalisation and metabolism in important crop species such as barley and rice, and energy yield in grasses with rapid growth and high biomass such as the African elephant grass *Miscanthus* that are considered as possible biofuels (Leonelli 2016, ch. 6). The ability to build on the extensive blueprint provided by model organism work, and related knowledge and tools, has undoubtedly accelerated discovery in other species and paved the way towards a systematic, integrative approach of to the study of biodiversity.

In this sense, the current 'reframing' of model organism research as reference for a much wider comparative exercise is not a reframing at all, but rather the culmination of this programme of biological work and the reasoning and resources underpinning it. The increase in opportunities for comparative approaches across species is not necessarily taking attention away from model organism research, but it is shifting perceptions of its role within biology (e.g., see Hedges 2002). Rather than serving as stand-ins for other organisms, model organisms are now explicitly used as baseline for cross-species comparison. Comparative tools are becoming part and parcel of what it means for a model organism to represent both whole organisms and other organisms. These roles would not have been possible without support from the various components of the model organism repertoire.

7.3 Towards New Approaches to Natural Variation

Comparative tools, and related big genotypic and phenotypic data, facilitate the identification of both differences and similarities among species. Emphasis on difference has enabled a systematic investigation of both significant and minute forms and sources of biological variation, including variation among model organism specimens and strains. This line of research provides a way to address one of the limitations of the ability of model organisms to 'represent', namely, the high degree of standardisation of the species and strains used and the well-recognised idealisations inherent in making their cultivation in laboratory settings more efficient and hence their use more tractable (Ankeny 2009; Love & Trevisano 2013). As we saw in Section 2, the processes of abstracting organisms for laboratory use involve purposefully departing from some aspects of the representational target, while emphasising others. It is well-recognised that specimens stocked in centralised collections and kept in highly rarefied laboratory settings do not represent how biology works in the natural world in any straightforward way, once organisms are in contact with an ever-changing and varying environment. In addition, over time the lab processes used to maintain stocks are likely to have allowed differences between model organisms and their wild relatives to accumulate, while at the same time lab-based stocks have of course been subject to evolutionary change.

Therefore, more recently there has been strong interest in assessing limitations of the commonly used strains of model organisms. For instance, in the case of the nematode *C. elegans*, experiments have documented selective pressures that appear to have led to the fixation of laboratory-derived alleles for particular genes in the typical research strain (Bristol or N2, see Ankeny 2001), which have influenced a large number of traits as well as behaviours that in turn affect experimental interpretations (Sterken et al. 2015). In addition, strong phenotypic effects caused by these laboratory-derived alleles are claimed to be hindering the discovery of 'natural' alleles. Hence with *C. elegans* and other model organisms, there has been renewed attention to comparing results across different lab environments, including comparisons related to nutrition, living quarters, breeding processes, and the extent to which they interact with other organisms in the laboratory environment including microbes.

These efforts have been accompanied by more focus on documenting and studying natural variants of various model organism species around the world and comparing stock specimens with 'wild' specimens. For example, the recently established *Caenorhabditis elegans* Natural Diversity Resource (Cook et al. 2017) aims to isolate and collect wild strains from a variety of natural environments. It provides genome-wide sequence and variant data for every strain as well as integrated tools for comparative analyses in order to support study of how populations of individuals are genetically different from one another and how those differences might impact disease, particularly given the medical and agricultural importance of nematodes.

These types of efforts have been described as aiming to develop a 'natural history of model organisms' particularly given we know very little about the 'real lives' of many model organisms (e.g., Alfred & Baldwin 2015). Additional examples include increased focus on wild house mouse (*M. musculus*) populations in terms of the spread of their resistance to rodenticides. This research focus permits investigation both of various adaptation processes given the commensal lifestyle of humans and mice, but also molecular and biochemical explorations of specific anticoagulant mechanisms in this context in comparison to the standard lab strains (for a review, see Phifer-Rixey & Nachman 2015). Both lab strains and 'natural' populations of various *Drosophila* species have been used comparatively to study responses to changing environments, particularly because certain species have very limited climatic distributions; hence they provide ways to study the genetic bases of adaptations to extreme climates and potentially to address problems of species loss in the face of global warming and other anthropogenic climatic changes (Hoffmann 2010; Markow 2015). These types of uses are yet another form of comparative research, which allow model organisms (and their close relations) to be useful in new domains and wider set of fields, especially in medicine, and ecology and environmental science.

In terms of our account, we can see in these processes a conscious weakening in the commitments of researchers in terms of their previous focus on highly standardised organisms in fully standardised environments, which were key components of the (original) model organism repertoire. As argued throughout this Element, during much of the history of model organism research, environmental variation became something to be disregarded since it introduced far too much complexity to an already very complex experimental problem, that of understanding organisms as integrated wholes. As understanding of model organisms increased over the years within that delimited context, and the repertoires around model organisms became more complex and sophisticated, researchers have acquired the ability to actively reconsider the experimental environments in which organisms are kept. They are thus able to assess the extents to which those set ups and their underlying commitments affect their research, as also occurred in the case of rodents as models for alcoholism discussed in Section 6 (Ankeny et al. 2014).

7.4 The Digital, the Biological, and the Synthetic

These emerging approaches have broader implications for how biological work is done, and particularly efforts to digitise organismal research and the data associated with it. As we have argued, a fundamental component of the model organism repertoire was the building of large-scale accessible databases incorporating genomic and other types of data for reuse both by those working with these organisms but also permitting applications beyond those communities including comparative use. However, there are well-recognised limitations to what has been collected: many databases do not have much systematic information (in other words, metadata) on some of the broader aspects of such research, for example, on the environments in which experiments were carried out (Leonelli 2016).

In these cases, it is likely that the information provided in data infrastructures is being supplemented with and interpreted through working assumptions about the uniformity of laboratory environments and techniques employed by different investigators. The regular meetings among model organism researchers, for instance, at dedicated conferences and workshops, also no doubt facilitated the sharing of background knowledge and training in relevant methods and techniques. Continued reliance on these mechanisms is however problematic in light of the increasing size of the communities as well as the growing recognition of the many factors that can influence experimental results (as discussed in Section 7.3). These issues have deep implications. For instance, it may be problematic to reuse existing data when they are gathered using mutants grown in highly insular laboratory environments that were not necessarily standardised or whose characteristics were not recorded in any detail and hence are not retrievable as metadata. In turn, the parameters used to collect and organise data on model organisms are shifting. Additional research will be required to update and supplement what has been collected in order to be able to reuse and integrate data on various organisms in a productive manner.

These issues point to the continued importance of carrying out new experiments (or reproducing old ones) with established model organisms, particularly at a time when various conceptual components associated with the model organism repertoire have come into flux, and in light of the changing roles of these models in the wider research landscape. The significance of the relation between analogue and digital work on model organisms is underscored by long-standing attempts to ensure and retain easily retrievable links between data and the material samples on which data were originally obtained, as exemplified for instance, within the *Arabidopsis* community. This close link between the material and the data produced from them can be viewed as a cautionary note in response to hyped interpretations of the revolutionary import of artificial intelligence and work on simulated, digital organisms for biological understanding (Leonelli 2018). As much as these tools are indeed transforming the research world, they do not make wet laboratory work with actual biological materials superfluous, not least of all because of the importance of verifying the quality and reliability of information through engagement with actual organisms.

Model organisms also are crucial to research in synthetic biology and associated attempts to recreate or fundamentally modify living organisms to fit human aims, precisely because they facilitate the application of computational tools and sophisticated engineering techniques using organisms that are highly predictable, easily obtainable, and better understood than any other species. Synthetic biology arguably was been built upon a very small set of model organisms, primarily *E. coli* and *S. cerevisiae*, and it has recently been recognised that it is necessary to move to 'next generation' chassis about which there is a large knowledge base and adequate technologies, among other factors, but critically which will allow applications that can be deployed in the field to be developed, in part by looking to organisms beyond the lab (e.g., Adams 2016). As argued, for instance, in the latest comprehensive review on the past and future of *Arabidopsis* research (Provart et al. 2016), its role in supporting synthetic plant biology constitutes its most attractive feature: the integrative understanding acquired on *A. thaliana* underpins efforts to engineer plant networks in heterologous contexts and engineer novel networks with genetic components from other organisms.

Thus, to underscore what has sometimes been called the 'paradox of model organisms' (Hunter 2008), the need for and reliance on them is likely to diminish only when most of the fundamental biological processes have been detailed which in turn will permit greater use of other organisms including human biomaterials as well as in silico and digital methods.

7.5 Conclusion: Situating Organisms as Models

In their original form, model organisms were not models of organisms in their natural environments but entities separated from their natural environments and explored in standardised laboratory conditions which in turn functioned as anchors for a broad and ever-evolving set of modelling ecosystems. They were able to serve these functions in the context of a range of laboratory conditions, methods, tools, and infrastructures. They were situated within particular communities that shared conceptual commitments and experimental methods, as well as fitting with a broader landscape which had certain political, social, and institutional features. These features all come together in the model organism repertoire.

Model organisms are thus necessarily what we term *situated* models. Repertoires play a critical role in shaping the interpretations that are required whenever a model is taken as representing, that is, when properties exemplified by models are associated with properties of their representational targets. Repeated use of and reference to similar organisms kept under the same general environmental conditions has provided considerable opportunities for sharing knowledge across a vast constellation of biological disciplines, groups, and research schools. These processes have made model organisms into movable resources that can be easily resituated. We contend that model organisms are in the process of being shifted yet again to a range of different roles. In particular, they are being used for comparative and integrative investigations that take the role of environment as something to be studied rather than held constant, and for synthetic and digital approaches to engineering life.

Bibliography

Adams, Bryn L. 2016. 'The Next Generation of Synthetic Biology Chassis: Moving Synthetic Biology from the Laboratory to the Field'. *ACS Synthetic Biology* 5: 1328–30.

Alfred, Jane and Ian T. Baldwin. 2015. 'The Natural History of Model Organisms: New Opportunities at the Wild Frontier'. *eLife* 4: e06956.

Ankeny, Rachel A. 2000. 'Fashioning Descriptive Models in Biology: Of Worms and Wiring Diagrams'. *Philosophy of Science* 67: S260–72.

Ankeny, Rachel A. 2001. 'The Natural History of *Caenorhabditis Elegans* Research'. *Nature Reviews Genetics* 2: 474–8.

Ankeny, Rachel. 2009. 'Model Organisms as Fictions'. In *Fiction in Science: Philosophical Essay on Modeling and Idealization*, edited by Mauricio Suárez, 193–204. London: Routledge.

Ankeny, Rachel A. 2010. 'Historiographic Reflections on Model Organisms: Or How the Mureaucracy May Be Limiting Our Understanding of Contemporary Genetics and Genomics'. *History and Philosophy of the Life Sciences* 32: 91–104.

Ankeny, Rachel A. and Sabina Leonelli. 2011. 'What's So Special about Model Organisms?' *Studies in History and Philosophy of Science* 42: 313–23.

Ankeny, Rachel A. and Sabina Leonelli. 2015. 'Valuing Data in Postgenomic Biology: How Data Donation and Curation Practices Challenge the Scientific Publication System'. In *Postgenomics: Perspectives on Biology after the Genome*, edited by Hallam Stevens and Sarah Richards, 126–49. Durham: Duke University Press.

Ankeny, Rachel A. and Sabina Leonelli. 2016. 'Repertoires: A Post-Kuhnian Perspective on Scientific Change and Collaborative Research'. *Studies in History and Philosophy of Science* 60: 18–28.

Ankeny, Rachel A. and Sabina Leonelli. 2018. 'Organisms in Experimental Research'. In *The Historiography of Biology*, edited by Michael Dietrich, Mark Borrello, and Oran Harman, 1–25. Dordrecht: Springer.

Ankeny, Rachel A. and Sabina Leonelli. 2020. 'Using Repertoires to Explore Changing Practices in Recent Coral Research'. In *From the Beach to the Bench: Why Marine Biological Studies?* edited by Karl Matlin, Jane Maienschein, and Rachel A. Ankeny, 249–70. Chicago: University of Chicago Press.

Ankeny, Rachel A., Sabina Leonelli, Nicole C. Nelson, and Edmund Ramsden. 2014. 'Making Organisms Model Human Behavior: Situated Models in North-American Alcohol Research, since 1950'. *Science in Context* 27: 485–509.

Baetu, Tudor. 2016. 'The "Big Picture": The Problem of Extrapolation in Basic Research'. *British Journal for the Philosophy of Science* 67: 941–64.

Bangham, Jenny. 2019. 'Living Collections: Care and Curation at Drosophila Stock Centres'. *BJHS Themes* 4: 123–47.

Bevan, Michael and Sean Walsh. 2004. 'Positioning Arabidopsis in Plant Biology: A Key Step toward Unification of Plant Research'. *Perspectives on Translational Biology* 135: 602–06.

Blum, Martin and Tim Ott. 2019. 'Xenopus: An Undervalued Model Organism to Study and Model Human Genetic Disease'. *Cells Tissues Organs* 205: 303–12.

Bolker, Jessica A. 1995. 'Model Systems in Developmental Biology'. *BioEssays* 17: 451–55.

Bolker, Jessica A. 2009. 'Exemplary and Surrogate Models: Two Modes of Representation in Biology'. *Perspectives in Biology and Medicine* 52: 485–99.

Bolker, Jessica A. and Rudolf A. Raff. 1997. 'Beyond Worms, Flies and Mice: It's Time to Widen the Scope of Developmental Biology'. *Journal of NIH Research* 9: 35–39.

Brigandt, Ingo and Alan Love. 2017. 'Reductionism in Biology'. In *The Stanford Encyclopedia of Philosophy*, edited by Edward N. Zalta, https://plato.stanford.edu/cgi-bin/encyclopedia/archinfo.cgi?entry=reduction-biology.

Burian, Richard M. 1993. 'How the Choice of Experimental Organism Matters: Epistemological Reflections on an Aspect of Biological Practice'. *Journal of the History of Biology* 26: 351–67.

Burian, Richard M. 1997. 'Exploratory Experimentation and the Role of Histochemical Techniques in the Work of Jean Brachet, 1938–1952'. *History and Philosophy of the Life Sciences* 19: 27–45.

Callon, Michel. 1984. 'Some Elements of a Sociology of Translation: Domestication of the Scallops and the Fishermen of St Brieuc Bay'. *The Sociological Review* 32: 196–233.

C. elegans Sequencing Consortium. 1998. 'Genome Sequence of the Nematode *C. Elegans*: A Platform for Investigating Biology'. *Science* 282: 2012–18.

Cannatella, David C. and Rafael O. De Sá. 1993. '*Xenopus Laevis* as a Model Organism'. *Systematic Biology* 42: 476–507.

Caporael, Linnda R., James R. Griesemer, and William C. Wimsatt. 2014. 'Developing Scaffolds: An Introduction', In *Developing Scaffolds in*

Evolution, Cognition, and Culture, edited by Linnda R. Caporael, James R. Griesemer, and William C. Wimsatt, 1–20. Cambridge: MIT Press.

Cartwright, Nancy. 1989. *Nature's Capacities and Their Measurement*. Oxford: Oxford University Press.

Chang, Hasok. 2012. *Is Water H_2O? Evidence, Pluralism and Realism*. Dordrecht: Springer.

Cicero, Theodore J. 1979. 'Critique of Animal Analogues of Alcoholism'. In *Biochemistry and Pharmacology of Ethanol*, vol. 2, edited by Edward Majchrowicz and Ernest P. Noble, 533–60. New York: Plenum.

Cold Spring Harbor Protocols. 2019. *Emerging Model Organisms*, http://cshprotocols.cshlp.org/site/emo/.

Cook-Deegan, Robert. 1994. *The Gene Wars: Science, Politics, and the Human Genome*. New York: W.W. Norton & Company.

Cook, Daniel E., Stefan Zdraljevic, Joshua P. Roberts, and Erik C. Andersen. 2017. 'CeNDR, The Caenorhabditis Elegans Natural Diversity Resource'. Nucleic Acids Research 45: D650–7.

Craver, Carl F. and Lindley Darden. 2013. *In Search of Mechanisms: Discoveries across the Life Sciences*. Chicago: University of Chicago Press.

Davies, Gail. 2013. 'Arguably Big Biology: Sociology, Spatiality and the Knockout Mouse Project'. *BioSocieties* 8: 417–31.

Davies, Jamie A. 2007. 'Developmental Biologists' Choice of Subjects Approximates to a Power Law, with No Evidence for the Existence of a Special Group of "Model Organisms"'. *BMC Developmental Biology* 7: 40–46.

Davis, Rowland H. 2004. 'The Age of Model Organisms'. *Nature Reviews Genetics* 5: 69–76.

de Chadaravian, Soraya. 2002. *Designs for Life: Molecular Biology after World War II*. Cambridge: Cambridge University Press.

de Chadaravian, Soraya and Nick Hopwood, editors. 2004. *Models: The Third Dimension of Science*. Stanford: Stanford University Press.

Dietrich, Michael R., Rachel A. Ankeny, and Patrick M. Chen. 2014. 'Publication Trends in Model Organism Research'. *Genetics* 198: 787–94.

Dietrich, Michael R., Rachel A. Ankeny, Nathan Crowe, Sara Green, and Sabina Leonelli. 2020. 'How to Choose Your Research Organism'. *Studies in History and Philosophy of the Biological and Biomedical Sciences* 80: 101227.

Frigg, Roman and Stephan Hartmann. 2018. 'Models in Science'. In *The Stanford Encyclopedia of Philosophy*, edited by Edward N. Zalta, https://plato.stanford.edu/archives/sum2018/entries/models-science/.

Frigg, Roman and James Nguyen. 2018. 'The Turn of the Valve: Representing with Material Models'. *European Journal for Philosophy of Science* 8: 205–44.

Gan, Xiangchao, Oliver Stegle, Jonas Behr, Joshua G. Steffen, Philipp Drewe, Katie L. Hildebrand, Rune Lyngsoe, et al. 2011. 'Multiple Reference Genomes and Transcriptomes for *Arabidopsis Thaliana*'. *Nature* 477: 419–23.

Gelfert, Axel. 2016. *How to Do Science with Models: A Philosophical Primer.* Dordrecht: Springer.

Germain, Pierre Luc. 2014. 'From Replica to Instruments: Animal Models in Biomedical Research'. *History and Philosophy of the Life Sciences* 36: 114–28.

Gest, Howard. 1995. 'Arabidopsis to Zebrafish: A Commentary on "Rosetta Stone" Model Systems in the Biological Sciences'. *Perspectives in Biology and Medicine* 39: 77–85.

Gilbert, Scott F. 2001. 'Ecological Developmental Biology: Developmental Biology Meets the Real World'. *Developmental Biology* 233: 1–12.

Gilbert, Scott F. 2009. 'The Adequacy of Model Systems for Evo-Devo: Modeling the Formation of Organisms/Modeling the Formation of Society'. In *Mapping the Future of Biology*, edited by Anouk Barberousse, Michel Morange, and Thomas Pradeu, 57–68. Dordrecht: Springer.

Green, Sara. 2013. 'When One Model Is Not Enough: Combining Epistemic Tools in Systems Biology'. *Studies in History and Philosophy of Biological and Biomedical Sciences* 44: 170–80.

Green, Sara, Michael R. Dietrich, Sabina Leonelli, and Rachel A. Ankeny. 2018. '"Extreme" Organisms and the Problem of Generalization: Interpreting the Krogh Principle'. *History and Philosophy of the Life Sciences* 40: 1–22.

Griesemer, James. 1990. 'Material Models in Biology'. *PSA: Proceedings of the Biennial Meeting of the Philosophy of Science Association* 2: 79–93.

Griesemer, James. 2000. 'Development, Culture, and the Units of Inheritance'. *Philosophy of Science* 67: S348–68.

Griesemer, James. 2004. 'Three-Dimensional Models in Philosophical Perspective'. In *Models: The Third Dimension of Science*, edited by Soraya de Chadarevian and Nick Hopwood, 433–42. Stanford: Stanford University Press.

Griesemer, James R. 2013. 'Integration of Approaches in David Wake's Model-Taxon Research Platform for Evolutionary Morphology'. *Studies in History and Philosophy of Biological and Biomedical Sciences* 44: 525–36.

Grunwald, David J. and Judith S. Eisen. 2002. 'Headwaters of the Zebrafish: Emergence of a New Model Vertebrate'. *Nature Reviews Genetics* 3: 717–24.

Güttinger, Stephen. 2019. 'A New Account of Replication in the Experimental Life Sciences'. *Philosophy of Science* 86: 453–71.

Hacking, Ian. 1983. *Representing and Intervening: Introductory Topics in the Philosophy of Natural Science*. Cambridge: Cambridge University Press.

Hedges, S. Blair. 2002. 'The Origin and Evolution of Model Organisms'. *Nature Review Genetics* 3: 838–49.

Hickford, Daniel, Stephen Frankenberg, and Marilyn B. Renfree. 2009. 'The Tammar Wallaby, *Macropus eugenii*: A Model Kangaroo for the Study of Developmental and Reproductive Biology'. *Cold Spring Harbor Protocols* 2009(12): pdb.emo137.

Hilgartner, Stephen. 1995. 'Biomolecular Databases: New Communication Regimes for Biology?' *Science Communication* 17: 240–63.

Hilgartner, Stephen. 2017. *Reordering Life: Knowledge and Control in the Genomics Revolution*. Cambridge: MIT Press.

Hoffmann, Ary A. 2010. 'Physiological Climatic Limits in *Drosophila*: Patterns and Implications'. *Journal of Experimental Biology* 213: 870–80.

Hoffer, L. John. 2003. 'Complementary or Alternative Medicine: The Need for Plausibility'. *Canadian Medical Association Journal* 168: 180–82.

Hopwood, Nick. 2011. 'Approaches and Species in the History of Vertebrate Embryology. In *Vertebrate Embryogenesis: Embryological, Cellular and Genetic Methods*, edited by Francisco J. Pelegri, 1–20. New York: Humana Press.

Huber, Lara and Lara K. Keuck. 2013. 'Mutant Mice: Experimental Organisms as Materialised Models in Biomedicine'. *Studies in History and Philosophy of Biological and Biomedical Sciences* 44: 385–91.

Hunter, Philip. 2008. 'The Paradox of Model Organisms'. *EMBO Reports* 9: 717–20.

Jenner, Ronald A. and Matthew A. Wills. 2007. 'The Choice of Model Organisms in Evo-Devo'. *Nature Reviews Genetics* 8: 311–9.

Jones, Kathryn Maxson, Rachel A. Ankeny, and Robert Cook-Deegan. 2018. 'The Bermuda Triangle: The Pragmatics, Policies, and Principles for Data Sharing in the History of the Human Genome Project'. *Journal of the History of Biology* 51: 693–805.

Jørgenson, C. Barker. 2001. 'August Krogh and Claude Bernard on Basic Principles in Experimental Physiology'. *BioScience* 51: 59–61.

Keller, Evelyn Fox. 2000. 'Models of and Models for: Theory and Practice in Contemporary Biology'. *Philosophy of Science* 67: S72–86.

Kelty, Christopher M. 2012. 'This Is Not an Article: Model Organism Newsletters and the Question of "Open Science"'. *BioSocieties* 7: 140–68.

Kimmel, Charles B. 1989. 'Genetics and Early Development of Zebrafish'. *Trends in Genetics* 5: 283–88.

Knuuttila, Tarja. 2011. 'Modeling and Representing: An Artefactual Approach'. *Studies in History and Philosophy of Science* 42: 262–71.

Kohler, Robert E. 1994. *Lords of the Fly: Drosophila Genetics and the Experimental Life.* Chicago: University of Chicago Press.

Krebs, Hans A. 1975. 'The August Krogh Principle: "For Many Problems There Is an Animal on Which It Can Be Most Conveniently Studied"'. *Journal of Experimental Zoology* 194: 221–26.

Krogh, August. 1929. 'Progress in Physiology'. *American Journal of Physiology* 90: 243–51.

Laubichler, Manfred D. and Gerd B. Müller, editors. 2007. *Modeling Biology: Structures, Behaviors, Evolution. Vienna Series in Theoretical Biology.* Cambridge: MIT Press.

Leonelli, Sabina. 2007a. 'What is in a Model? Using Theoretical and Material Models to Develop Intelligible Theories'. In *Modeling Biology: Structures, Behaviors, Evolution. Vienna Series in Theoretical Biology*, edited by Manfred D. Laubichler and Gerd B. Müller, 15–36. Cambridge: MIT Press.

Leonelli, Sabina. 2007b. 'Growing Weed, Producing Knowledge: An Epistemic History of *Arabidopsis thaliana*'. *History and Philosophy of the Life Sciences* 29: 55–87.

Leonelli, Sabina. 2008. 'Performing Abstraction: Two Ways of Modelling *Arabidopsis Thaliana*'. *Biology & Philosophy* 23: 509–23.

Leonelli, Sabina. 2009. 'The Impure Nature of Biological Knowledge'. In *Scientific Understanding: Philosophical Perspectives*, edited by Henk de Regt, Sabina Leonelli, and Kai Eigner, 189–209. Pittsburgh: Pittsburgh University Press.

Leonelli, Sabina. 2010. 'Packaging Small Facts for Re-Use: Databases in Model Organism Biology'. In *How Well Do Facts Travel?: The Dissemination of Reliable Knowledge*, edited by Peter Howlett and Mary S. Morgan, 325–48. Cambridge: Cambridge University Press.

Leonelli, Sabina. 2013. 'Global Data for Local Science: Assessing the Scale of Data Infrastructures in Biological and Biomedical Research'. *BioSocieties* 8: 449–65.

Leonelli, Sabina. 2016. *Data-Centric Biology: A Philosophical Study.* Chicago: University of Chicago Press.

Leonelli, Sabina. 2018. 'Re-Thinking Reproducibility as a Criterion for Research Quality'. *Research in the History of Economic Thought and Methodology: Including a Symposium on the Work of Mary Morgan: Curiosity, Imagination, and Surprise* 36B: 129–46.

Leonelli, Sabina and Rachel A. Ankeny. 2012. 'Re-Thinking Organisms: The Impact of Databases on Model Organism Biology'. *Studies in History and Philosophy of Biological and Biomedical Sciences* 43: 29–36.

Levy, Arnon and Adrian Currie. 2015. 'Model Organisms Are Not (Theoretical) Models'. *British Journal for the Philosophy of Science* 66: 327–48.

Lewis, Jamie, Paul Atkinson, Jean Harrington, and Katie Featherstone. 2013. 'Representation and Practical Accomplishment in the Laboratory: When Is an Animal Model Good-Enough?' *Sociology* 47: 776–92.

Love, Alan C. 2008. 'Explaining Evolutionary Innovation and Novelty: Criteria of Explanatory Adequacy and Epistemological Prerequisites'. *Philosophy of Science* 75: 874–86.

Love, Alan C. and Michael Travisano. 2013. 'Microbes Modeling Ontogeny'. *Biology & Philosophy* 28: 161–88.

Maher, Brendan. 2009. 'Biology's Next Top Model?' *Nature* 458: 695–98.

Markow, Theresa A. 2015. 'The Natural History of Model Organisms: The Secret Lives of Drosophila Flies'. *eLife* 4: e06793.

Meunier, Robert. 2012. 'Stages in the Development of a Model Organism as a Platform for Mechanistic Models in Developmental Biology: Zebrafish, 1970–2000'. *Studies in History and Philosophy of Biological and Biomedical Sciences* 43: 522–31.

Minelli, Alessandro and Jan Baedke. 2014. 'Model Organisms in Evo-Devo: Promises and Pitfalls of the Comparative Approach'. *History and Philosophy of the Life Sciences* 36: 42–59.

Mitchell, Sandra D. 2003. *Biological Complexity and Integrative Pluralism*. Cambridge: Cambridge University Press.

Morgan, Mary S. 2003. 'Experiments without Material Intervention: Model Experiments, Virtual Experiments and Virtually Experiments'. In *The Philosophy of Scientific Experimentation*, edited by Hans Radder, 216–35. Pittsburgh: University of Pittsburgh Press.

Morgan, Mary S. 2007. 'Reflections on Exemplary Narratives, Cases, and Model Organisms'. In *Science Without Laws: Model Systems, Cases, Exemplary Narratives*, edited by Angela N. H. Creager, M. Norton Wise, and Elizabeth Lunbeck, 264–74. Durham: Duke University Press.

Morgan, Mary S. 2012. *The World in the Model: How Economists Work and Think*. Cambridge: Cambridge University Press.

Morrison, Margaret and Mary S. Morgan. 1999. 'Models as Mediating Instruments'. In *Models as Mediators: Perspectives on Natural and Social Science*, edited by Margaret Morrison and Mary S. Morgan, 10–37. Cambridge: Cambridge University Press.

Mungall, Christopher J., David B. Emmert, and The FlyBase Consortium. 2007. 'A Chado Case Study: An Ontology-Based Modular Schema for Representing Genome-Associated Biological Information'. *Bioinformatics* 23: i337–46.

Mungall, Chistopher J., Nicole L. Washington, Jeremy Nguyen-Xuan, Christopher Condit, Damian Smedley, Sebastian Köhler, et al. 2015. 'Use of Model Organism and Disease Databases to Support Matchmaking for Human Disease Gene Discovery'. *Human Mutation* 36: 979–84.

National Institutes of Health (NIH). Model organisms for biomedical research. www.nih.gov/science/models/.

Nelson, Nicole C. 2018. *Model Behavior: Animal Experiments, Complexity, and the Genetics of Psychiatric Disorders*. Chicago: University of Chicago Press.

Nicholson, Daniel J. and John Dupré, editors. 2018. *Everything Flows: Towards a Processual Philosophy of Biology*. Oxford: Oxford University Press.

O'Malley, Maureen A. 2007. 'Exploratory Experimentation and Scientific Practice: Metagenomics and the Proteorhodopsin'. *History and Philosophy of the Life Sciences* 29: 337–60.

O'Malley, Maureen A., Ingo Brigandt, Alan C. Love, John W. Crawford, Jack A. Gilbert, Rob Knight, et al. 2014. 'Multilevel Research Strategies and Biological Systems'. *Philosophy of Science* 81: 811–28.

Parkkinen, Veli Pekka. 2017. 'Are Model Organisms Theoretical Models?' *Disputatio* 9: 471–98.

Perlman, Robert L. 2016. 'Mouse Models of Human Disease: An Evolutionary Perspective'. *Evolution, Medicine, and Public Health* 2016: 170–76.

Phifer-Rixey, Megan and Michael W. Nachman. 2015. 'The Natural History of Model Organisms: Insights into Mammalian Biology from the Wild House Mouse *Mus musculus*'. *eLife* 4: e05959.

Pickstone, John. 2001. *Ways of Knowing: A New History of Science, Technology, and Medicine*. Chicago: University of Chicago Press.

Piotrowska, Monika. 2013. 'From Humanized Mice to Human Disease: Guiding Extrapolation from Model to Target'. *Biology and Philosophy* 28: 439–55.

Provart, Nicholas J. et al. 2016. '50 Years of Arabidopsis Research: Highlights and Future Directions'. *New Phytologist* 209: 921–44.

Rader, Karen A. 2004. *Making Mice: Standardizing Animals for American Biomedical Research, 1900–1955*. Princeton: Princeton University Press.

Ramsden, Edmund. 2015. 'Making Animals Alcoholic: Shifting Laboratory Models of Addiction'. *Journal of the History of the Behavioral Sciences* 51: 164–94.

Ransohoff, Richard M. 2018. 'All (Animal) Models (of Neurodegeneration) Are Wrong. Are They Also Useful?' *The Journal of Experimental Medicine* 215: 2955–58.

Ratti, Emanuele. 2018. '"Models of" and "Models for": On the Relation between Mechanistic Models and Experimental Strategies in Molecular Biology'. *The British Journal for the Philosophy of Science*, axy018, https://doi.org/10.1093/bjps/axy018

Rhee, Seung Y. 2004. 'Carpe Diem: Retooling the "Publish or Perish" Model into the "Share and Survive" Model'. *Plant Physiology* 134: 543–47.

Rheinberger, Hans-Jörg. 1997. *Toward a History of Epistemic Things: Synthesizing Proteins in the Test Tube*. Stanford: Stanford University Press.

Rheinberger, Hans-Jörg. 2010. *An Epistemology of the Concrete: Twentieth-Century Histories of Life*. Durham: Duke University Press.

Rheinberger, Hans-Jörg and Staffan Müller-Wille. 2010. *A Cultural History of Heredity*. Chicago: University of Chicago Press.

Robert, Jason Scott. 2008. 'The Comparative Biology of Human Nature'. *Philosophical Psychology* 21: 425–36.

Rosenberg, Alexander. 1996. 'The Human Genome Project: Research Tactics and Economic Strategies'. *Social Philosophy and Policy* 13: 1–17.

Rosenthal, Nadia and Michael Ashburner. 2002. 'Taking Stock of Our Models: The Function and Future of Stock Centres'. *Nature Reviews Genetics* 3: 711–17.

Rosenthal, Nadia and Steve Brown. 2007. 'The Mouse Ascending: Perspectives for Human-Disease Models'. *Nature Cell Biology* 9: 993–99.

Russell, James J., Julie A. Theriot, Pranidhi Sood, Wallace F. Marshall, Laura F. Landweber, Lillian Fritz-Laylin, et al. 2017. 'Non-Model Model Organisms'. *BMC Biology* 15: 1–31.

Schaffner, Kenneth F. 1986. 'Exemplar Reasoning about Biological Models and Diseases: A Relation between the Philosophy of Medicine and Philosophy of Science'. *Journal of Philosophy and Medicine* 11: 63–80.

Schaffner, Kenneth F. 1998. 'Model Organisms and Behavioral Genetics: A Rejoinder'. *Philosophy of Science* 65: 276–88.

Schaffner, Kenneth F. 2001. 'Extrapolation from Animal Models: Social Life, Sex, and Super Models'. In *Theory and Method in the Neurosciences*, edited by Peter K. Machamer, Rick Grush, and Peter McLaughlin, 200–30. Pittsburgh: University of Pittsburgh Press.

Schaffner, Kenneth F. 2016. *Behaving: What's Genetic, What's Not, and Why Should We Care?* New York: Oxford University Press.

Slack, Jonathan M.W. 2009. 'Emerging Market Organisms'. *Science* 323: 1674–75.

Smith, Richard N., Jelena Aleksic, Daniela Butano, Adrian Carr, Sergio Contrino, Fengyuan Hu, et al. 2012. 'InterMine: A Flexible Data Warehouse System for the Integration and Analysis of Heterogeneous Biological Data'. *Bioinformatics* 1: 3163–65.

Sommer, Ralf J. 2009. 'The Future of Evo-Devo: Model Systems and Evolutionary Theory'. *Nature* 10: 416–22.

Srinivasan, Balaji S., Nigam H. Shah, Jason A. Flannick, Eduardo Abeliuk, Antal F. Novak, and Serafim Batzoglou. 2007. 'Current Progress in Network Research: Toward Reference Networks for Key Model Organisms'. *Briefings in Bioinformatics* 8: 318–32.

Steel, Daniel. 2008. *Across the Boundaries: Extrapolation in Biology and Social Science*. Oxford: Oxford University Press.

Sterken, Mark G., L. Basten Snoek, Jan E. Kammenga, and Erik Christian Andersen. 2015. 'The Laboratory Domestication of Caenorhabditis elegans'. *Trends in Genetics* 31: 224–31.

Strasser, Bruno J. and Soraya de Chadarevian. 2011. 'The Comparative and the Exemplary: Revisiting the Early History of Molecular Biology'. *History of Science* 49: 317–36.

Sundberg, J. P., J. M. Ward, and P. Schofield. 2009. 'Where's the Mouse Info?' *Veterinary Pathology* 46: 1241–44.

Valenzano, Dario R., Aziz Aboobaker, Andrei Seluanov, and Vera Gorbunova. 2017. 'Non-Canonical Aging Model Systems and Why We Need Them'. *The EMBO Journal* 36: 959–63.

Waters, C. Kenneth. 2007. 'The Nature and Context of Exploratory Experimentation: An Introduction to Three Case Studies of Exploratory Research'. *History and Philosophy of the Life Sciences* 29: 275–84.

Waters, C. Kenneth. 2013. 'Molecular Genetics'. In *The Stanford Encyclopedia of Philosophy* edited by Edward N. Zalta (ed.), plato.stanford.edu/archives/fall2013/entries/molecular-genetics/.

Weber, Marcel. 2005. *Philosophy of Experimental Biology*. Cambridge: Cambridge University Press.

Weber, Marcel. 2007. 'Redesigning the Fruit Fly: The Molecularization of Drosophila'. In *Science without Laws: Model Systems, Cases, Exemplary Narratives*, edited by Angela N. H. Creager, Elizabeth Lunbeck, and M. Norton Wise, 23–45. Durham: Duke University Press.

Weisberg, Michael. 2007. 'Who is a Modeler?' *British Journal for the Philosophy of Science* 58: 207ch33.

Weisberg, Michael. 2013. *Simulation and Similarity*. New York: Oxford University Press.

Wimsatt, William. 2000. 'Generative Entrenchment and the Developmental Systems Approach to Evolutionary Processes'. In *Cycles of Contingency*, edited by Susan Oyama, Paul E. Griffiths, and Russell D. Grey, 219–37. Cambridge: MIT Press.

Wood, William, and the Community of *C. elegans* Researchers, editors. 1988. *The Nematode Caenorhabditis elegans*. New York: Cold Spring Harbor Laboratory.

Woody, Andrea I. 2000. 'Putting Quantum Mechanics to Work in Chemistry: The Power of Diagrammatic Representation'. *Philosophy of Science* 67: S612–27.

Acknowledgments

The contents of this Element were presented and discussed at numerous talks over the last decade. We are very grateful to hundreds of colleagues in the humanities, social sciences, and life sciences who gave us feedback and prodded us on.

We are indebted to Michael Dietrich, Roman Frigg, James Nguyen, and a very sharp and helpful anonymous reviewer for their detailed comments on the manuscript draft. For specific and substantive input through long-term discussions, we thank Dick Burian, Alberto Cambrosio, Jim Collins, Bob Cook-Deegan, Gail Davies, Elihu Gerson, Scott Gilbert, Sara Green, James Griesemer, Kathryn Maxson Jones, Alan Love, Jane Maienschein, Mary Morgan, Nicole Nelson, Edmund Ramsden, Hans-Jörg Rheinberger, and Jason Robert. The Exeter Centre for the Study of the Life Sciences hosted us while writing the manuscript, and we are grateful to our colleagues there, particularly John Dupré, as well as to administrator Chee Wong for her expert assistance. Sabina is also grateful to the GARNet committee for collaboration and relevant discussions over the last decade. The manuscript was expertly shepherded through production at CUP by Michael Ruse and Grant Ramsey. We are grateful for the proofreading provided by Laura Ruggles, and the support and input from her and the other Organisms team members, Karina Burns and Dook Shepherd.

This research was supported by the Australian Research Council via the Discovery Project "Organisms and Us: How Living Things Help Us to Understand Our World" (DP160102989) and the European Research Council under award number 335925.

Our deepest gratitude goes to our families, who endured our absences while on writing retreats and supported us in all possible ways (short of writing the manuscript themselves!): thank you Glenn, Luca, Michel, Leonardo, and Luna!

Cambridge Elements ≡

Elements in the Philosophy of Biology

Grant Ramsey
KU Leuven

Grant Ramsey is a BOFZAP research professor at the Institute of Philosophy, KU Leuven, Belgium. His work centers on philosophical problems at the foundation of evolutionary biology. He has been awarded the Popper Prize twice for his work in this area. He also publishes in the philosophy of animal behavior, human nature and the moral emotions. He runs the Ramsey Lab (theramseylab.org), a highly collaborative research group focused on issues in the philosophy of the life sciences.

Michael Ruse
Florida State University

Michael Ruse is the Lucyle T. Werkmeister Professor of Philosophy and the Director of the Program in the History and Philosophy of Science at Florida State University. He is Professor Emeritus at the University of Guelph, in Ontario, Canada. He is a former Guggenheim fellow and Gifford lecturer. He is the author or editor of over sixty books, most recently *Darwinism as Religion: What Literature Tells Us about Evolution; On Purpose; The Problem of War: Darwinism, Christianity, and their Battle to Understand Human Conflict;* and *A Meaning to Life.*

About the Series

This Cambridge Elements series provides concise and structured introductions to all of the central topics in the philosophy of biology. Contributors to the series are cutting-edge researchers who offer balanced, comprehensive coverage of multiple perspectives, while also developing new ideas and arguments from a unique viewpoint.

Cambridge Elements$^{\equiv}$

Philosophy of Biology

Printed in the United States
By Bookmasters